Biotechnological Advances, Phytochemical Analysis and Ethnomedical Implications of *Sapindus* species

Reetika Singh • Bechan Sharma

Biotechnological Advances, Phytochemical Analysis and Ethnomedical Implications of *Sapindus* species

Springer

Reetika Singh
Department of Biochemistry
University of Allahabad
Allahabad, Uttar Pradesh, India

Bechan Sharma
Department of Biochemistry
University of Allahabad
Allahabad, Uttar Pradesh, India

ISBN 978-981-32-9188-1 ISBN 978-981-32-9189-8 (eBook)
https://doi.org/10.1007/978-981-32-9189-8

This Springer imprint is published by the registered company Springer Nature Singapore Pte Ltd.
The registered company address is: 152 Beach Road, #21-01/04 Gateway East, Singapore 189721, Singapore

Preface

Since prehistoric era, trees are a fundamental part of human life. Life could not be imagining on Earth without the plants. Exploitation, unorganized cultivation, rapid deforestation, urbanization and high rate of industrialization create various degrees of threats for plants especially the tree species. Sustainable cultivation, development and conservation are required to save these trees and increase the forest areas. Plant biotechnology can perform foremost role for the rapid propagation and conservation of these important plant species. In spite of conventional propagation and all plant tissue culture techniques, somatic embryogenesis has played an imperative role in large-scale development of the prevailing stocks of germplasm as well as for their ex situ conservation of woody trees. In vitro conservation techniques overwhelmed the limitations of long regeneration cycle of woody trees and obviate the dependence on flowering and seed set. Plant tissue culture is the technique which serves as the platform for different innovative areas of applied plant science, plant biotechnology and agriculture. The technique can be successfully used for large-scale propagation of plants, production of haploids, scaling up of production of secondary metabolites, production of disease-resistant plants, inducing variability and conservation of germplasm. In addition, regeneration of whole plants from genetically engineered cells also needs tissue culture methods. The use of modern in vitro techniques on woody trees has opened new possibilities for rapid mass multiplication of the existing stocks of germplasm as well as for their ex situ conservation. Besides affording multiplication in limited time and space, the technique of tissue culture has overcome the limitations posed by long regeneration cycle of woody trees and dependence on seasons. Plant tissue culture techniques allow mass multiplication of plants from single explant (small pieces of the mother plant) in short period. In vitro material would be proved as the best source of important phytochemicals. These phytochemicals possess various pharmacological properties.

Genus *Sapindus* is a member of Sapindaceae family. This tree is commonly known as reetha, washnut and soapnut tree. It is known for their high economical and medicinal significance. The fruits of this tree contain high amount of saponins and are used as natural detergent for washing of hair and clothes since olden times. Every part of this plant possesses various pharmacological properties and is frequently used for treatment of various disorders.

Chapter 1 describes the general introduction of this book. In Chap. 2, the morphological characteristics of genus *Sapindus* have been described. Chapter 3 comprises the all traditional uses and ethanomedicinal values of the *Sapindus* species. In Chap. 4, biotechnological advances in *Sapindus* species have been explained with all the details, techniques and process of plant tissue culture. Regeneration through organogenesis and somatic embryogenesis is mentioned in details. Genetic fidelity through DNA-based molecular markers, namely RAPD and ISSR, has been also summarized in this chapter. Chapter 5 summarizes the pharmaceutical aspects and phytochemistry of *Sapindus* tree. This chapter also contains phytochemical evaluation from leaf, fruit, callus and in vitro leaf extracts. To evaluate the antioxidant capacity, DPPH assay, β-carotene/linoleic acid bleaching assay, reducing potential assay and lipid peroxidation assay were used and found to be significant antioxidant potential. Chapter 6 describes the pharmacological activities of this plant species. Chapter 7 explains the nanotechnological applications of *Sapindus* tree.

This book will be beneficial for botanists, taxonomists, graduates, scientists and policy-makers. Critical suggestions will be always welcome for the improvement of this book.

Allahabad, Uttar Pradesh, India

Reetika Singh
Bechan Sharma

Contents

About the Authors

Reetika Singh, currently working as DST-SERB Scientist, has completed her PhD from Banaras Hindu University, Varanasi, in Biotechnology. She has cleared several national level competitive exams like CSIR-JRF, ASRB-NET and GATE and secured higher rank. She was awarded with a prestigious National Postdoctoral Fellowship (N-PDF) of DST-SERB, New Delhi, in 2016. She has visited China as Indian Delegate in the 2nd BRICS Young Scientist Forum 2017. She has published six technical research papers and one review article in reputed international journals with high impact factors. She has also published eight book chapters in different edited books of various international publishers. She has attended several national and international symposia/conferences and presented her research findings in many scientific meetings. She has been acting as a Reviewer and an Editor for couple of reputed international journals. She is also an active member of two reputed scientific societies in India.

Bechan Sharma is presently working as a Professor and Ex-Head of the Department of Biochemistry Allahabad University, Allahabad. He received his BSc (Honours), MSc and PhD in Biochemistry from BHU, Varanasi, in 1980, 1982 and 1988, respectively. The areas of his research interest include molecular biology of HIV/AIDS, tropical diseases (filariasis/malaria), drug design and development and biochemical toxicology. He has received number of awards/honours and successfully completed numerous important academic/administrative assignments. He possesses about 30 years of teaching and research experience, has carried out several research projects and has published over 200 research papers including book chapters and molecular methods in peer-reviewed international and national journals of repute. He has one US patent on HIV-1 genome structure-based anti-HIV-1 drugs design to his credit. He has supervised 18 PhDs and 6 PDFs. He is member/ life member of several national/international scientific societies and attended numerous symposia/conferences in India and abroad. He is Chief Editor/Associate Editor/ Executive Editor and Member of Editorial Board of over 130 peer-reviewed international and national journals. He has been acting as Honorary Reviewer for over 150 international/national scientific journals. His four books have been published by international publishers, and two are in print. He has worked as a Visiting Scientist in the USA for over 3 years and visited different labs at Italy, France, Iran, Thailand, Hong Kong, Japan, Germany and Brazil to conduct different collaborative research projects related to molecular virology and drug resistance. He has been awarded with ICMR's Senior Scientist Fellowship 2014–2015 in Biomedical Sciences under Indo-USA joint research programme to work at NIEHS-NIH, NC, USA, with Prof. Samuel Wilson on DNA repair. Recently, he has been included as a member in the World Society for Virology, Sweden, and in an Indo-Japanese research initiative towards eradication of HIV/AIDS in India. He is a Fellow of Academy of Environmental Biology (AEB) and Bioved Research Society (BRS).

Department of Biochemistry, University of Allahabad, Allahabad, Uttar Pradesh, India

Abbreviations/Acronyms

ABA	Abscisic acid
ANOVA	Analysis of variance
BAP	6- benzylaminopurine
°C	Degree Celsius
cm	Centimetre (s)
2, 4- D	2, 4-Dichlorophenoxyacetic acid
DDW	Double distilled water
DPPH	1,1-Diphenyl-2-picrylhydrazyl
EDTA	Ethylenediaminetetraacetic acid
FE	Frequency of embryogenesis
g	Gram (s)
GAE	Gallic acid equivalent
h	Hour (s)
HCl	Hydrochloric acid
H_2O	Water
IAA	Indole acetic acid
IBA	Indole butyric acid
IC_{50}	50% inhibition capacity
IE	Intensity of embryogenesis
IVL	In vitro leaf
KIN	Kinetin
l	Litre (s)
LSD	Least significant test
M	Molar
MHA	Mueller-Hinton agar
µg	Microgram
µl	Microliter
µM	Micromole
mg	Milligram (s)
min	Minute
ml	Millilitre
mm	Millimetre
MS	Murashige and Skoog (1962)

N	Normality
NaOH	Sodium hydroxide
nm	Nanometre
P	Probability
PE	Plant extract
PGR	Plant growth regulator (s)
%	Percent
RE	Rutin equivalent
ROS	Reactive oxygen species
rpm	Revolution per minute
RP	Reducing power
s	Second
TPC	Total phenolic content
TFC	Total flavonoid content
TDZ	Thidiazuron
ZN	Zeatin
v	Volume
w	Weight
WPM	Woody plant medium

List of Figures

List of Tables

Introduction

1

Trees have been a vital component of biodiversity and an integral part of human life since historical days. Forest trees are renewable sources of food, fodder, fibre, timber, fuel wood and other valuable products such as herbal medicines, rubber, etc. Rapid deforestation and exhaustion of genetic resources together with escalating human needs play an important role in reduction of the forest cover from the earth's surface. The reduction in forest trees and depletion of biodiversity is an alarming threat for the environmental changes. Two-third part of total global photosynthesis is accounted from the agricultural plants and trees especially the forest species. About 50% of the tropical forests have already been destroyed. In India, forest cover is disappearing at an annual rate 1.5 mha/year. Presently, only 8% of land area is covered by the forest as against a mandatory 33% of the geographical area. A number of valuable medicinal plants are under the boundary of extinction. According to Thomas (1997), *The Red Data Book of India* has 427 entries of endangered species of which 28 are considered extinct, 124 endangered, 81 vulnerable, 100 rare and 34 insufficiently known species. Therefore, there is a great necessity to conserve tree ecosystems for both their environmental and aesthetic values. To maintain and sustain forest vegetation, the conventional method of propagation is not fulfilling the increasing demands of the present days. Some conventional approaches have been exploited for propagation and improvement, but tree breeding efforts are restricted to the most valuable and fast-growing species.

Successful application of biotechnological techniques provides the new avenue for large-scale propagation and improvement of elite characters of tree species. In vitro regeneration techniques using organogenesis and somatic embryogenesis have been established in several tree species with successful field transfer and high survival rate. Plant tissue culture techniques allow mass multiplication of plants from single explant (small pieces of the mother plant) in short period. Genetic fidelity testing through molecular markers proves that the consistency in the elite parental characters in their progenies. Genetic engineering techniques open a new era in the biotechnology. In addition, regeneration of whole plants from genetically

R. Singh, B. Sharma, *Biotechnological Advances, Phytochemical Analysis and Ethnomedical Implications of Sapindus species*,
https://doi.org/10.1007/978-981-32-9189-8_1

engineered cells also needs tissue culture methods. The use of modern in vitro techniques on woody trees has opened new possibilities for rapid mass multiplication of the existing stocks of germplasm as well as for their ex situ conservation. Besides affording multiplication in limited time and space, the technique of tissue culture has overcome the limitations posed by long regeneration cycle of woody trees and dependence on seasons.

Nature was the first solution of each need and problem. Forest trees have provided mankind with a dynamic natural laboratory not only as sources of medicine, cosmetics and food but also as an essential element in the stabilization of the ecosystem (Schuster 2002). Ethnomedicine and ethnopharmacology are as old as human civilization. In India, Ayurveda has been in existence since ancient times. At that time, plant was only the option to cure the various ailments of peoples. Peoples lived around nature, and plants were like the family members. Ancestors correlated the mythology with plants for the conservation of plants.

Plants have always been exploited as the richest source of drugs of traditional systems of medicine, modern medicines, nutraceuticals, food supplements, folk medicines, pharmaceutical intermediates and chemical entities for synthetic drugs (Hammer et al. 1999). Over 50% of all allopathic drugs are of natural products (Stiffness and Douros 1982), and the natural product plays an important role in drug development programmes of the pharmaceutical industry (Baker et al. 1995; Cordell 1995). Clinical microbiologists have great interest in screening of medicinal plants for antimicrobial activity and phytochemicals as potential new therapeutics. The principles of many drugs found in plants are secondary metabolites (Ghani 2003). Dobelis (1993) such as alkaloids, flavanoids and terpenoids, which are responsible for antimicrobial activity (Jack and Okorosaye 2008). The development of drug resistance in human pathogens against commonly used antibiotics has compelled us for new antimicrobial substances from other sources including plants (Erdogrul 2002). Therefore the preliminary screening of medicinal plants for antimicrobial activity and other phytochemicals can establish a platform for further development on the research. Only a small percentage of medicinal plants traded in India are solely cultivated. The obligatory demand for a huge raw material of medicinal plants is met from their wild populations. Over-exploitation and/or destructive harvesting to meet such demand in fact threatens the survival of many rare species (Natesh 1999). Confronted by such unprecedented genetic erosion and disappearance of species and ecosystems, conservation of natural resources assumes paramount urgency. Pharmaceutical development from botanical resources in the forest areas offers great potential for contributing to sustained growth. Consequently, there is an urgent need to conserve plants especially the forests as biological resources in order to ensure the future availability of known and yet undiscovered pharmaceutical and herbal products for coming generations.

Around 250,000 higher plant species are present on earth, and approximately 80,000 plant species are medicinal. India is one of the world's 12 biodiversity centres having more than 45,000 different plant species. The plant biodiversity of India is matchless due to the presence of 16 different agro-climatic zones, 10 vegetation zones, 25 biotic provinces and 426 biomes. In India, tropical and subtropical regions are richer in biodiversity.

Among all families, Sapindaceae is one of the important families of plant kingdom consisting of about 140–150 genera and 1500–2000 species (Ferrucci 2000). Earlier this family was known as Lychee family. Genus *Sapindus* has about 5–12 species. These are either shrubs or small trees naturally occurring in the warm temperate to tropical regions. *Sapindus* includes both deciduous and evergreen species widespread over Himalayan range. Most of the plants of Sapindaceae family are known for their medicinal or economical values. *Sapindus* species is an important medicinal and economical plant. Various biotechnological tools and techniques such as in vitro micropropagation, molecular markers, clonal fidelity testing, etc. have been used for the germplasm conservation of this tree. Genus *Sapindus* is known for pharmaceutically active compound saponins that are present in the all species of *Sapindus*. Different kinds of saponins are present in the fruits of this plant. Saponins are responsible for various pharmacological properties such as antimicrobial activity, immunomodulatory, hepatoprotective activity, anticancer activity, etc. There is an urgent need to compile all the literature about this plant at one place. This book will be beneficial to study all the aspects of this plant.

Questions

1. Which part of this plant contains the high amount of saponins?
2. What factors are playing an important role in destruction of forests?
3. Genus *Sapindus* belongs to which plant family?
4. How many genus and species belong to the *Sapindus* plant family?

Further Readings

Singh, R. (2015). PhD thesis, Banaras Hindu University, Varanasi, UP, India.

Upadhyay, A., & Singh, D. K. (2012). Pharmacological effects of *Sapindus mukorossi*. *Revista do Instituto de Medicina Tropical de São Paulo, 54*, 273–280.

Morphological Characteristics of *Sapindus* Species

2

The plants belonging to the Sapindaceae family are one of the most important forest species to be conserved and valued in Africa due to their multiple uses, high nutritional content and medicinal value. These plants exist as trees and shrubs and tendril-bearing vines with about 140–150 genera and 1400–2000 species worldwide (Adeyemi 2011). The plants of Sapindaceae family are geographically distributed in the temperate and tropical regions of the world. The majority of species are native to Asia, although there are a few in South America, Africa and Australia (APG II 2003). In Africa continent, only two species of *Sapindus* named *S. saponaria* and *S. trifoliatus* are present (Adeyemi and Ogundipe 2012).

2.1 Botanical Classification of *Sapindus*

Kingdom: Plantae
(Unranked): Angiosperms
(Unranked): Eudicots
Class: Magnoliopsida
Subclass: Rosidae
Order: Sapindales
Family: Sapindaceae
Subfamily: Sapindoideae
Genus: *Sapindus*

Members of the genus *Sapindus* are commonly known as soapberries or soapnuts as the fruit pulp is used to make soap. The generic name *Sapindus* is derived from the Latin words *saponis*, which means "soap", and *indicus*, which means "of India". In India, *Sapindus* includes two major species *S. mukorossi* and *S. emarginatus* in North and South India. Many other species of *Sapindus* have also been investigated for their astringent and pharmacological properties.

R. Singh, B. Sharma, *Biotechnological Advances, Phytochemical Analysis and Ethnomedical Implications of Sapindus species*,
https://doi.org/10.1007/978-981-32-9189-8_2

Table 2.1 Different species of *Sapindus* and their geographical distribution

S. no.	Plant name	Common name	Geographical distribution/native place
1	*Sapindus trifoliatus*	Indian soapberry, ritha	India
2	*Sapindus saponaria*	Western soapberry	America, Florida, Texas, Arizona
3	*Sapindus mukorossi*	Chinese soapberry	Northern India, east to the Himalayas but native of China
4	*Sapindus delavayi*	Not known	China, India
5	*Sapindus detergens*	Soapnut, ritha	
6	*Sapindus emarginatus*	Not known	Southern Asia
7	*Sapindus laurifolia*	Ritha	India
8	*Sapindus marginatus*	Florida soapberry	Florida to South Carolina
9	*Sapindus vitiensis*	Not known	America, Samoa, Fiji
10	*Sapindus tomentosus*	Not known	China
11	*Sapindus oahuensis*	Lonomea	Hawaii
12	*Sapindus rarak*	Not known	Southeast Asia

Genus *Sapindus* comprises several species and is distributed over the world. *Sapindus* species occupies a significant place of high medicinal and economical value due to being a rich source of saponins. *Sapindus* is a genus of about 5–12 species of shrubs and small trees (family: Sapindaceae). This family is commonly known as Lychee family. This is native to warm temperate to tropical regions in both the Old World and New World. Members of the genus *Sapindus* are commonly known as soapberries or soapnuts because the fruit pulp is used to make soap. The number of species is not fixed, and it is disputed between the different authors. In North America one to three species are accepted. A number of *Sapindus* species are known, but scientists have the controversy on the few species. *S. saponaria* var. saponaria is considered as *S. marginatus* and is included in by some workers. Some workers have also the contradiction on *S. laurifolia* and *S. trifoliatus*. A summary of all the known species of this plant is mentioned in Table 2.1.

Of these mentioned species, only three species named *S. mukorossi*, *S. trifoliatus* and *S. saponaria* are common and frequently explored for research and commonly discussed in the literature. Very little work has been done on *S. emarginatus* and *S. laurifolia*. All over the world, scientists focused more on these three species for their research. Among these *S. mukorossi* is more explored for phytochemical and pharmacological purposes. We concluded all the aspects in details of these three important species in the book.

Sapindus species (Family-Sapindaceae) is a medium-sized to large, deciduous tree native to warm temperate to tropical regions of the world and commonly cultivated in peninsular India. The tree is a native of South India where it is common on the coasts and in the open forests at low elevations. It is also occasionally cultivated around villages in West Bengal, Bihar, Madhya Pradesh and Uttar Pradesh. Common names include soapberry and soapnut, both names referring to the use of the crushed fruits to make soap. Soapnuts contain saponin, a natural detergent which can be used to clean clothes and hairs. Soapnuts have become popular as an alternative to laundry detergent among those who have opted for an eco-friendly lifestyle. All the species have some characteristic morphological features. These morphological features are discussed in separately for each species.

2.2 *Sapindus mukorossi*

Sapindus mukorossi Gaertn. is a deciduous tree, distributed mainly in tropical and subtropical regions of Asia (Fig. 2.1). This plant is native to China and Japan, but it is also remarkably scattered to Singapore and India. *S. mukorossi* is commonly known as Chinese soapberry as China is its native place. In India, this tree is also

Fig. 2.1 Morphology of different parts of *Sapindus mukorossi* tree. (**a**) *S. mukorossi* tree with fresh flush of leaves, (**b**) tree with mature leaves, (**c**) a single leaf with leaflets, (**d**) inflorescence, (**e**) a single flower, (**f**) a twig with flowers and fruits

recognized by numerous other names such as soapnut, soapberry, washnut, reetha, aritha, dodan, dodani, etc. In Japan and China, its fruits are used for the cure of various health disorders. In Japan, its pericarp is called "enmei-hi", which means "life prolonging pericarp" and in China "wu-huan-zi", the "non-illness fruit" (Tanaka et al. 1996). This tree is very popular due to its great economical and medicinal values.

S. mukorossi is a deciduous tree that grows well in deep clay loamy soil with an annual rainfall of 150–200 mm (Goyal et al. 2014), and this tree is widely distributed in upper reaches of Indo-Gangetic plains, Shivaliks and sub-Himalayan region at altitudes from 200 to 1500 m. It is commonly present in some regions of North India like Jammu-Kashmir, Himachal Pradesh, Haryana Uttar Pradesh and Uttarakhand. *S. mukorossi* is a fairly large tree usually grown up to height of 12–15 m, but sometimes it attains a height 18–20 m with 1.8 m girth. The bark of this plant is dark to pale yellow, fairly smooth, having many vertical lines of lenticels and fine fissures exfoliating in irregular wood scales. Leaves are 30–50 cm long, alternate, paripinnate and having 5–10 pairs leaflets, opposite or alternate (Fig. 2.1).

Leaflets are lanceolate, acuminate and 2.5–5 cm in length (Orwa et al. 2009). The appearance of new leaves starts in March, and from April to December, the trees are covered with leaves. Leaves turn yellow in the end of December and start to fall in December–January. The tree remains leafless till March when new leaves appear. Inflorescences are compound terminal panicles about 30 cm long with pubescent branches. Flowers are about 5 mm across, polygamous, greenish white, subsessile, numerous, mostly bisexual with five sepals. The panicles of white or purplish flowers appear in April. Fruiting starts in the month of May and becomes mature in June–July. Ripening of fruits occurs in October–November, and fruit colour changes from yellow orange to dark brown during ripening. Fruits are globose, fleshy, usually one-seeded drupe, sometimes two drupels together, about 1.8–2.5 cm across. Seeds are 0.8–1.3 cm in diameter, globose, smooth, black and loosely present in dry fruit (Fig. 2.2) (Orwa et al. 2009).

2.3 *Sapindus trifoliatus*

Sapindus trifoliatus is also known as soapnut tree, a deciduous tree, mainly distributed in peninsular India, has very high economic value in saponin industry and is valued for its important medicinal properties (Fig. 2.3). Like the *S. mukorossi* its fruits contain saponins and are mostly used as detergent. Saponins are a natural detergent and are mainly used for washing cloth and hair as substitute of soap. Root bark are also saponaceous and used as a natural detergent, in the same way as the fruits (Anonymous 1972; Kirtikar and Basu 1999). A fruit's pericarp contains the fair amount of saponins and sugars (Anonymous 1972; Gupta and Ahmed 1990). Peoples used its fruits to manufacture soap and as remedies against asthma, diarrhoea and cholera (Anonymous 1972). The saponin moiety is characterized as hederagenin group of glycosides (Kanchanapoom et al. 2001).

Fig. 2.2 Morphological features of *S. mukorossi* fruits. (**a**) Early coming fruits with three lobes, (**b**) mature fruits, (**c**) seeds of mature fruits, (**d**) mature ripe fruits

The *S. trifoliatus* occurs in the Western Ghats and plains of South India, so it is also known as South Indian soapberry. This is a deciduous tree and flourishes well in deep clay loamy soil with an annual rain fall of 200 mm. This tree may attain the height of 25 m. Leaves are compound, alternate, pinnate and 15–30 cm long. The leaflets are elliptic-lance shaped, smooth with tipped points, slightly oblique based, terminal pair longest and terminal leaflet often absent. Leaflets are almost stalkless, mostly present in 2–3 pairs and are 8–18 cm long, and the width is 5–7.5 cm. Flowering starts in November to January every year. The flowers form in large panicles. Flowers are greenish white, in terminal, slightly velvety panicles. Flower stalks are 3 mm long, velvety. Sepals are five, slightly fused at the base, 4–5 mm long, ovate-oblong, velvety. Petals are five in number, free, 5–6 mm long, lance-shaped to ovate, clawed and bristly. Disc is five-lobed. Stamens are eight, free, filaments 2–3 mm long. Ovary is three locular, three-lobed, ovoid, about 3 mm long, velvety and with one ovule in each locule. The fruits are solitary globose and appear in the month of July–August. The fruit is a small leathery-skinned drupe 1–2 cm in diameter. Fruit is two- to three-lobed, 1.3–2 cm long, velvety when young, hard and smooth when mature. Seed is round, black, 0.6–0.9 cm in diameter. The fruits and seeds are slightly smaller than the North Indian species (www.flowersofindia.com; Kommalapati et al. 1998; Meena et al. 2012; Rao and Lakshmi 2012).

Fig. 2.3 Morphology of *Sapindus trifoliatus* tree. (**a**) *S. trifoliatus* tree, (**b**) leaf, (**c**) mature fruits, (**d**) ripe fruits. (Source: Asthana 2010, PhD Thesis; www.flowersofindia.com)

2.4 *Sapindus saponaria*

Sapindus saponaria is a very popular plant of genus *Sapindus*, native in American regions and local populations, and has been used as traditional and folk medicine since olden periods (Fig. 2.3). This tree is commonly distributed from Missouri to Louisiana in the southwest and from Oklahoma and Texas to southern Colorado, New Mexico, Southern Arizona, Northern Mexico and Brazilian Amazon region in western region (Vora 1989; Albiero et al. 2001). This plant is commonly known as wingleaf soapberry, western soapberry and jaboncillo (USDA 2010; NRCS Plants database, USDA 2015).

S. saponaria is a small to medium deciduous tree, 7.5–15.5 m tall. The branches are very dense in comparison to the *S. mukorossi* and *S. trifoliatus*. Leaves are pinnately compound, thick and leathery. The leaves are 20–38 cm in length; 6–20 leaflets are present having smooth margin and long tapered tips (Wikipedia 2018). This plant is described as polygamodioecious. That is, individual trees in a population may be truly dioecious (having only male or female flowers), or they may contain flowers with both male and female functions. The flowers are small, white in colour, borne in rather large clusters of terminal or axillary panicles, open during May to July. The fruits are yellow, translucent, globular drupe measuring 1.0–1.4 cm in diameter. The fruits usually contain a single, dark brown, hard-coated seed, but sometimes two or three seeds may be present. Ripening of fruits occurs during September to October and persists on the tree until late winter or spring (Munson 1984; Khatamian and Abuelgasim 1986) (Fig. 2.4).

S. saponaria has two subspecies named *S. saponaria* var. *saponaria* and *S. saponaria* var. *drummondii*. These two varieties of S. saponaria are morphologically different from each other. These differences are summarized in Table 2.2.

2.5 *Sapindus rarak*

Sapindus rarak is also a species of soapberry plant. It is a tall tree, native of Southeast Asia, and is now widely distributed in all over Asia. Its species name comes from the Malaysian name rerak or rerek (Nianhe and Gadek 2012; Slik 2012). In China, it's local name is Maobanwuhuanzi. S. rarak is also known by other names viz. *Sapindus angustifolius*, *S. rarak* and *Dittelasma rarak*, but *S. rarak* is a most acceptable name of this *Sapindus* species.

S. rarak is a deciduous tree and grows up to 42 m in height, native to South and East Asia. It is distributed from India (Assam) and Sri Lanka in the west to south, China and Taiwan in the north and to Indonesia in the south. It generally grows in forests at 200–1600 m altitude. In Indonesia its common name is lerak or klerek (Fajriaty et al. 2014). Leaves compound, 7–9 in pair, up to 50 cm long and glabrous. Leaflets are subopposite to alternate, lanceolate-ovate, apex obtuse to tapering acute-acuminate, midrib slightly raised above. Inflorescences are up to 35 cm long, densely fulvous- to ferrugineous-tomentellous. Flowers are zygomorphic, white in colour,

Fig. 2.4 Morphological features of *Sapindus saponaria*. (**a**) tree, (**b**) leaf arrangement, (**c**) mature fruits, (d) ripen fruits. (Source: http://www.sbs.utexas.edu/bio406d/images/pics/sap/sapindus_saponaria.htm; https://www.etsy.com/au/listing/561566097/sapindus-saponaria-soap-berry-6-seeds; https://hort.ifas.ufl.edu/treesandpowerlines/sapindus_saponaria_spring.shtml)

sepals flat, long fulvous-hairy with four petals, lanceolate-ovate to elliptic. The fruit parts are subglobular, 2 × 1.8 cm in dimension, carinate, red, glabrous (Fig. 2.5).

Seeds subglobular, 1.2–1.5 cm diam. Fruits and seeds are used as buttons and beads, and wood is hard but not durable (Nianhe and Gadek 2012; Wikipedia 2018, www.efloras.org). *S. rarak* var. *rarak* and *S. rarak* var. *velutinus* are the two varieties of *S. rarak*. *S. rarak* var. *rarak* have the leaf axis, and abaxial surface of leaflets is glabrous, and in *S. rarak* var. *velutinus*, leaf axis and abaxial surface of leaflets are densely appressed villous (www.efloras.org).

Similar to the other soapberries, the fruits are rich in saponins and used as natural soap (Wina et al. 2003). Like the other *Sapindus* species, this plant has also been used for washing by the native Asians and Native Americans since thousands of

Table 2.2 Differences between the two varieties of *S. saponaria*

S. no.	Features	*S. saponaria* var. *saponaria*	*S. saponaria* var. *drummondii*
1	Trunk	Grey to reddish in colour and scaly bark	Light grey in colour, scaly with thin plate-like bark and sometimes shallowly furrowed
2	Twig	Grey in colour and hairless	Grey-brown in colour and hairy
3	Midveins of leaf	Midveins on leaves of var. *saponaria* are mostly winged	Those of var. *drummondii* are without winged
4	Flowering season	November	May–June
5	Buds	Small brown and hairless	Small dark brown and hairy
6	Fruit ripening season	Occurs in spring	Occurs in October

Source: https://en.wikipedia.org/wiki/Sapindus_saponaria

Fig. 2.5 Morphology of *Sapindus rarak* tree. (**a**) *S. rarak* tree, leaf and fruits (**b**) mature fruits, (Source: https://alchetron.com/Sapindus-rarak; https://www.google.co.in)

years (Stoffels 2008). Soapnuts are being used for commercial production of cosmetics, detergents and many more products. Unlike the other detergent, it does not cause the fading of colours (www.ryan-isra.net/lerak-kelerak-werek-rerek-lamuran-manfaat-kegunaan/).

2.6 Traditional Uses of *Sapindus Species*

Since ancient times, *Sapindus* sps. especially *S. trifoliatus* and *S. mukorossi* have been used as a washing detergent for silk and wool. The fruits of this plant were utilized by Indian jewellers for restoring the brightness of tarnished ornaments made of gold, silver and other precious metals (Singh et al. 2010). It was also used for washing and bleaching cardamoms (Anonymous 1992). This plant can also be used for afforestation, soil reclamation and phytoremediation (Dobhal et al. 2012). Traditionally, its fruits are being used for washing hair and clothes (Kirtikar and Basu 1999; Singh et al. 2010). The fruit is of considerable medicinal importance and is used for the treatment of epilepsy, pimples, migraines, eczema, psoriasis, etc. It also shows insecticidal activities and is used to remove lice from the scalp (Goyal et al. 2014). Powdered seeds are used for the treatment of dental caries, arthritis, common cold, constipation and nausea. In Ayurveda, the seeds are used to remove tan and freckles from the skin (Goyal et al. 2014). The leaves are used in baths to relieve from joint pain, and the roots are used to cure gout and rheumatism (Upadhyay and Singh 2012). The tree is used for silviculture practice and social forestry programmes (Troup 1921; Dev and Guha 1979). In Taiwan the tree is used for flatland reforestation programme (Shiau et al. 2009). Some of the traditional applications of the genus *Sapindus* are demonstrated in Table 2.3.

Out of several *Sapindus* species, *S. mukorossi*, *S. trifoliatus* and *S. saponaria* are more common and used for medicinal purposes due to easy occurrence. Only few research literatures are present about other species; therefore only these three species will be discussed in details in the next chapters.

Questions

1. Which *Sapindus* species are known as American soapberry?
2. Which *Sapindus* species are known as Chinese soapberry due to origin in China?
3. Which two *Sapindus* species are more common in India?
4. Which part of this plant contains high amount of saponin?
5. What is the most common use of this plant?

Table 2.3 Traditional uses of genus *Sapindus*

Parts	Uses/details	References
Seeds	Restoring the brightness of gold and silver metals	Singh et al. (2010)
Whole plant	Afforestation, soil reclamation and phytoremediation	Dobhal et al. (2012)
	Silviculture	Troup (1921)
	Social forestry	Dev and Guha (1979)
Fruit pericarp	Solubilizing property	Pelegrini et al. (2008) and Nakayama et al. (1986)
Fruits	Biodiesel production from oil	Chen et al. (2013)

6. What are the main morphological distinct features among the *S. trifoliatus*, *S. mukorossi* and *S. saponaria*?
7. How many species of Sapindus have been widely exploited for medicinal purposes?

Further Readings

Kirtikar, K. R, & Basu, B. D. (1999). *Indian medicinal plants 1* (pp. 632–635). New Delhi: International Book Distributors.

www.efloras.org, *Sapindus rarak* Candolle, Prodr. 1:608. 1824, Flora of China, 12:11, retrieved on 17/05/2018.

3 Traditional Uses and Ethnomedicinal Values of *Sapindus* spp.

Since human civilization life-saving drugs have been in demand. Ancient literature showed that barks of some specific trees and mud were used for the treatment of wounds in the Stone Age. Since the historical period, plants were the only source of primary health care due to easy availability and effectiveness. Due to the course of time, some advances also occurred in Ayurveda, the natural therapy. India and China have a rich traditional medicinal system. In India, beside the Ayurveda, homoeopathy, Unani and Sidha are also reliable medicine systems for the treatment of the various ailments and have been used for several decades. These are all the medicine systems based on the plant-based natural compounds.

Plethora of drugs, nutraceuticals, pharmaceuticals and pharmaceutical intermediates are present in the plants (Hammer et al. 1999). Recently, elite medicinal plants have gain the commitment of the pharmaceutical and scientific societies towards production of natural antioxidants. Production of natural antioxidants involved the several steps including isolation, characterization and identification of secondary metabolites and their implication as active principles in drug formulations (Taylor et al. 2001).

The emergence of new infectious, chronic and drug-resistant diseases has stimulated workers to explore the medicinal plants for a new therapeutics for prevention and cure. Besides the traditional approaches, biotechnology and molecular science advances have made it progressively realistic to refurbish the traditional medicines into contemporary drugs. These innovative methodologies also raise new concerns to explore the economic, ethical and policy considerations of drug development centred on indigenous knowledge.

R. Singh, B. Sharma, *Biotechnological Advances, Phytochemical Analysis and Ethnomedical Implications of Sapindus species*,
https://doi.org/10.1007/978-981-32-9189-8_3

3.1 History of Traditional Medicine

Ayurveda is a comprehensive medicine system, considered as the ancient traditional medicine system in India for more than 5000 years. Ayurveda was fully established around 2500–600 BC. Human body is composed of three body doshas, three mental doshas, dhatus and malas. Disbalance of these doshas causes several diseases. The disease management of illness requires balancing the doshas back into a normal state through various ways like lifestyle mediations, spiritual encouragement and dealing with herbo-mineral formulations.

In India approximately 15,000–20,000 medicinal plants have been reported. Out of these only 7000–7500 species are utilized for the herbal medicinal formulations by traditional groups. Ayurveda and Unani are Indian ancient medicine systems, which have the documentation of herbal origin drugs. Herbal drugs can be derived either from the whole plant or from different organs such as leaves, stem, bark, root, flower, seed, etc. Excretory plant products such as gum, resins and latex have also been used for the preparation of some drugs. Even the allopathic medicine system has accepted many plant-derived drugs which form an important segment of the modern pharmacopoeia. The plant-derived drug offers a stable market worldwide, but also plants continue to be an important source for new drugs.

During ancient civilization, India has been considered as amusing depository of medicinal plants. The forest area acts as the principal repository of huge number of medicinal and aromatic plants, which are generally collected as raw materials for production of drugs and perfumery products. In Ayurveda, approximate 8000 herbal remedies have been mentioned. The *Rigveda* (5000 BC) has recorded 67 medicinal plants, *Yajurveda* 81 species and *Atharvaveda* (4500–2500 BC) 290 species, and *Charaka Samhita* (700 BC) and *Sushruta Samhita* (200 BC) had described properties and uses of 1100 and 1270 species, respectively. These plants are still used in the classical herbal formulations in the Ayurvedic medicine system.

3.2 Resources of Medicinal Materials

Ethnomedicine and the history of medicine have in common the description of medical systems, whereby the separation is geographical in the former case and historical in the latter case. As experts in both fields are not at the same time practitioners of these medical systems, they depend on source material. The interpretation of these sources is largely subordinated to the subjective personal structure of the scientist, on the one hand, and is closely related to epistemological problems, coloured by the spirit of the times, on the other hand. Above all, many medical historians even today fall victim to an unjustified cultural evolutionism, according to which ethnomedical research work in the field of "primitive medicine" may be employed to reconstruct a fictive paleopathology. The cooperation which is nowadays established between ethnomedicine and the history of medicine concerns the common structures of medical systems in their social and cultural context – so-called pattern – with the aim of establishing a theory of medical thinking and behaviour of man (Prinz 1991).

3.3 Current Status at National and International Levels

Since olden time plants have been used as primary therapeutic agents for the cure of various diseases. Thousands of plants were documented due to their therapeutic values in the ancient literature. According to the estimation of the World Health Organization, about 80% of the world population especially in developing countries depends on herbal medicine for primary and basic health care (Vines 2004). The increasing toxicity and side effects associated with the unselective use of synthetic drugs and an antibiotic have improved the interest in the application of herbal drugs. Subsequent global preference towards herbal medicine has advanced the growth of plant-based pharma industries. The production of traditional medicine at global market was valued at US$ 83 billion annually in 2008; this rate will increase as exponential. This is likely to touch US$ 5 trillion by 2050 (Krishnan et al. 2011). Approximately 250,000 higher plant species are present on earth; out of these more than 80,000 are reported as medicinal. Out of the 80,000 medicinal plants, approximately 15,000–20,000 plants possess rich medicinal significance. However, only 7000–7500 plant species are utilized for their medicinal properties by traditional communities.

3.4 Validation of Traditional Medicinal Values Through Scientific Experiments

Green plants synthesize and preserve a variety of biochemical products, many of which are extractable and used as chemical feedstocks or as raw material for various scientific investigations. Many secondary metabolites of plant are commercially important and find use in a number of pharmaceutical compounds. Plants may provide biologically active molecules and lead structures to develop the modified derivatives with improved activity and/or reduced toxicity. Approximately 120 therapeutic agents of known structure have been identified and isolated from about 90 plant species of flowering plants. Some of the valuable plant drugs are vinblastine, vincristine, taxol, podophyllotoxin, camptothecin, digitoxigenin, gitoxigenin, digoxigenin, tubocurarine, morphine, codeine, aspirin, atropine, pilocarpine, capsaicin, allicin, curcumin, artemisinin and ephedrine among others. Sometimes the crude extracts of medicinal plants may be used for medication. In India only a small percentage of medicinal plants are solely cultivated. The requisite demand for a large raw material of medicinal plants is met from their wild populations. The published literature delivers state-of-the-art information on the most inventive science, research, industry, market and the future of ethnomedicine and drug discovery.

3.5 Herbal Shampoo Preparation

The fruits of *Sapindus* species have the potential of herbal soap. The peoples of rural area were using the fruit's pericarp for washing the hair for avoiding the dandruff and dryness. Mali et al. (2010) attempted the preparation of herbal shampoo from *S. mukorossi* along with *Asparagus racemosus* and *Acacia concinna* (Mali et al. 2010). They used a total of nine plants: *S. mukorossi*, *Asparagus racemosus* and *Acacia concinna* for isolation of saponins; *Glycyrrhiza glabra*, *Azadirachta indica*, *Hibiscus rosa-sinensis* and *Bassia malabarica* were utilized for making the base of shampoo; and *Citrus reticulate*, *Emblica officinalis* and *Aloe vera* were used for isolation of herbal juice. Selection of these plants was based on the extensive literature survey towards presence of chemical constituents and biological activities towards the hair and skin. After the preparation of the herbal base of shampoo, they added some chemical ingredients such as xanthan gum, glycerine, EDTA, methyl paraben and orange oil in different concentrations and finally prepared the herbal shampoo.

3.6 Traditional Medicinal Uses of *Sapindus* spp.

Sapindus species have the high traditional medicinal values and used for the time of Ayurveda. Pharmacological and therapeutic values of this plant are due to the occurrence of various phytochemicals in general and the saponins in particular. The fruit is of considerable medicinal importance and is used for the treatment of epilepsy, pimples, migraines, eczema, psoriasis, etc. The plants also possess insecticidal properties and are used to get rid of lice from the scalp (Goyal et al. 2014). The powdered seeds have been used for the management of arthritis, common cold, constipation, dental caries and nausea. Ayurvedic literature also documented the use of seeds for removal of tan and freckles from the skin (Goyal et al. 2014). Besides the fruits and seeds, the leaves and roots also possess the therapeutic properties and used to cure the joint pain such as gout and rheumatism, respectively (Upadhyay and Singh 2012). The ethnopharmacological and traditional uses of *Sapindus mukorossi* are enlisted in the Table 3.1.

Questions

1. Describe the traditional uses of different species of *Sapindus*.
2. Describe the ethnopharmacological uses of different species of *Sapindus*.
3. Mention the name of few plant-based natural compounds.
4. Explain the international status of herbal medicine.

Table 3.1 Ethnopharmacological and traditional uses of *Sapindus mukorossi*

Sapindus sps.	Parts	Ethnopharmacological/ traditional uses	References
S. mukorossi	Fruit pulp	Alternative of soap	Philomina (2010)
	Fruits	Herbal shampoo preparation	Mali et al. (2010)
	Fruit, leaf, seed and roots	Expectorant, source of natural surfactant insecticide, treating pimples, excessive salivation, dental caries, epilepsy, chlorosis, migraine, arthritis, common cold, constipation, nausea, joint pain, gout and rheumatism	Upadhyay and Singh (2012), Kasai et al. (1986), Singh et al. (2010), Dhar et al. (1989) and Tanaka et al. (1996)
	Not specified	Treatment of eczema and psoriasis and for removing freckle	Kirtikar and Basu (1991)
S. saponaria	Not specified	Treatment of ulcer, inflammation and external wounds	Vora (1989) and Albiero et al. (2001)
S. trifoliatus	Not specified	Used as tonic; used for the cure of diabetes, chronic dysentery, diarrhoea, cholera, stomachache and paralysis; astringent and sedative to uterus; used in eye's disorders; also used to induce labour pain and normal contraction of uterus after child birth	Meena et al. (2012), Sharma et al. (2011), Kishore et al. (2010), Rao and Lakshmi (2012), Arulmozhi et al. (2005a, b, c) and Krishnaveni and Thaakur (2008)
	Not specified	Used for the cure of headaches, toothaches, fever, coughs, bronchitis, arthritis, osteoarthritis, lower cholesterol and blood pressure and aid in preventing internal blood clots worldwide	Mahmood et al. (2013)
	Decoction of fruit pulp	For relief in hemicrania and for bringing back the consciousness during epileptic and hysteric attack	Kritikar and Basu (1999)
		Anti-spermatogenic and anti-androgenic activities and anti-rheumatic activities	Anonymous (1972), Bodhankar et al. (1974), Dixit and Gupta (1982) and Pandey and Sharma (1986)
		Anti-helminthic and used for the treatment of asthma	Grover et al. (2005)
	Roots and bark	Used as expectorant and demulcent	Asthana et al. (2011)

Further Readings

Anonymous. (1977). *The Ayurvedic formulary of India. 1. Part II.* New Delhi: Ministry of Health and Family Welfare, Government of India.

Anonymous. (1979). *Wealth of India: Raw materials. Part I.* New Delhi: CSIR.

Jaiswal. et al. (2016). Botanical drugs in Ayurveda and traditional Chinese medicine. *Journal of Ethnopharmacology, 194*, 245–259.

Biotechnological Advances in *Sapindus* sps. 4

4.1 Introduction of Biotechnology

The term biotechnology was used for the first time by a Hungarian engineer named Karl Ereky in 1917. The biotechnology was defined by various scientists in their own way. But in simple words, the term "biotechnology" can be defined as "the use of modern technology with combination of basic biological sciences for human welfare". One of the scientists defined that biotechnology is "the science of applied biological sciences" (Anonymous 1981). The origin of biotechnology is as old as human civilization. Therefore, biotechnology can be divided into two phases, (i) old (traditional) biotechnology and (ii) new (modern) biotechnology. The traditional biotechnology refers to the conventional technology which has been used for many centuries. Beer, wine, cheese and many foods have been produced using traditional biotechnology. Thus, the traditional biotechnology includes the process that is based on the natural methods. Modern biotechnology started from basic plant tissue culture (PTC) to molecular markers, genetic engineering, proteomic and transcriptomics to metabolomics.

The conventional method of propagation of trees limited with several inherent bottlenecks because trees are generally slow-growing, long-lived, sexually self-incompatible and highly heterozygous plants. Due to the prevalence of high heterozygosity in these species, a number of recessive deleterious alleles are retained within populations, resulting in high genetic load and inbreeding depression. This limits the use of traditional breeding methods such as selfing and backcrossing and makes it difficult to fix desirable alleles in a particular genetic background (Williams and Savolainen 1996). Thus conventional breeding is rather slow and less productive and inefficient for the genetic improvement of trees. To avoid these impairments, vegetative or clonal propagation has been arrayed for recovering dominant, additive, and epistatic genetic effects to select superior genotypes. Plant tissue culture and genetic transformation methods offer an imperative option for effective and large multiplication and improvement of trees within a limited time frame.

R. Singh, B. Sharma, *Biotechnological Advances, Phytochemical Analysis and Ethnomedical Implications of Sapindus species*,
https://doi.org/10.1007/978-981-32-9189-8_4

Plant tissue culture is a basic and promising technique of biotechnology. The development of biotechnology started with the invention of PTC. Plant tissue culture also serves as a valuable tool to study the cell physiology and metabolism of plants and is applicable for the different innovative areas of applied plant science, plant biotechnology and agriculture (Von-Arnold et al. 2002). This technique has been successfully applied for large-scale propagation of plants, production of haploids and production of disease-resistant plants, inducing variability and conservation of germplasm of elite and genotype of plants in very short time period. Besides the large-scale propagation, PTC is also used for the scaling-up of production of secondary metabolites using different technologies such as cell culture, suspension culture, hairy root culture, etc. Large-scale production of phytochemicals was performed using bioreactors. The most significant advantage offered by tissue culture is to produce a large number of plants from a single explant in a relatively short time period and space (Merkle and Dean 2000). For the large-scale production of in vitro plants, mainly two distinct pathways are recognized (a) organogenesis and (b) somatic embryogenesis (Litz and Gray 1992). These two pathways are successfully developed in thousands of plants, and few are in practice at the commercial level. Both are discussed in details.

4.2 In Vitro Regeneration of Woody Trees

In woody plant species, micropropagation started with the culture of cambial tissues of *Ulmus campestris* in early 1940s by Gautheret (Thorpe et al. 1991), yet the progress has been very slow because of several problems (Swartz et al. 1991; Ziv 1991). At that time the phenolic execution was the major constraint which caused browning and death of explants (Debergh and Read 1991). A high rate of culture contamination due to the endophytic microorganisms was another major problem and phenotypic variations following into instability during field growth (Zimmerman 1985). The first successful report of formation of adventitious buds in woody trees was achieved in *Ulmus campestris* using cambial tissue by Gautheret (Thorpe et al. 1991). In 1950 organogenesis (shoot proliferation in callus cultures) was first time reported in a woody gymnosperm *Sequoia sempervirens* by Ball. The true plantlets were achieved (via organogenesis) in *Populus tremuloides* (Winton 1970) and *Pinus palustris* (Sommer et al. 1975). Later on successful in vitro regeneration protocols have been established using organogenesis as well as somatic embryogenesis in several tropical trees (Litz and Jaiswal 1991). General steps of in vitro regeneration have been shown in Fig. 4.1.

Problems in regenerating cultures of several recalcitrant tree species have been overcome by emphasizing the modification in media to explant selection. Media manipulation for supporting optimum morphogenesis in tissue culture of woody species is generally based on the MS medium (Murashige and Skoog 1962). Composition of MS medium is mentioned in Tables 4.1 and 4.2. Several researchers modified the original media (MS media); these modified media are known as B5 medium, MT medium, Nitsch medium, Woody Plant medium, etc. In vitro

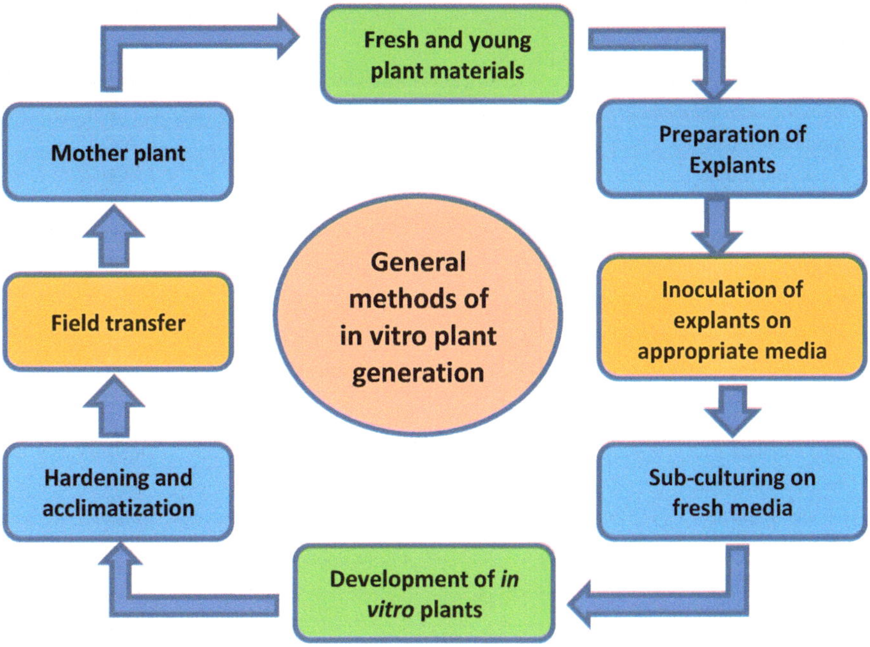

Fig. 4.1 General process for the development of in vitro plantlets

regeneration of woody plants can be influenced by the annual growth cycle of the tree and the seasonal variations (at the time of explant isolation) (Murashige 1974; Hammerschlag 1986). Seasonal variations significantly affect the regeneration potential of explants as well as percent survival. Effects of seasonal variations have been reported in *Eucalyptus*, *Psidium guajava*, *Bauhinia vahlii and Terminalia arjuna* (McComb and Bennett 1982; Amin and Jaiswal 1987; Dhar and Upreti 1999; Pandey et al. 2006).

4.3 Regeneration via Organogenesis

Nature of explant, composition of the medium and culture conditions are the leading factors that influence the method and rate of in vitro regeneration. Factors of the culture environment also influence the growth and differentiation of explants. These include (1) chemical composition of the nutrient culture medium, (2) pH of the medium, (3) plant growth regulators and other additives and (4) culture conditions (temperature and intensity, quality and duration of light exposed. Organogenesis and plant regeneration from seedling explants have been done in woody trees like *Terminalia chebula* (Shyamkumar et al. 2003), *Pterocarpus santalinus* (Rajeswari and Paliwal 2006), *Stereospermum personatum* (Shukla et al. 2009) and *Spondias mangifera* (Tripathi and Kumari 2010). Induction of organogenesis has been

Table 4.1 Composition of MS (Murashige and Skoog 1962) medium

Constituents	Chemical formula	Amount in stock (mg/l)	Amount in medium (mg/l)
Stock 1 (macronutrients)			
Ammonium nitrate	NH_4NO_3	33,000	1650
Potassium nitrate	KNO_3	38,000	1900
Calcium chloride	$CaCl_2.2H_2O$	8800	400
Magnesium sulphate	$MgSO_4.7H_2O$	7400	370
Potassium dihydrogen orthophosphate	KH_2PO_4	3400	170
Stock 2 (micronutrients)			
Potassium iodide	KI	166	0.83
Boric acid	H_3BO_3	1240	6.20
Manganese sulphate	$MnSO_4.4H_2O$	4460	22.30
Zinc sulphate	$ZnSO_4.7H_2O$	1720	8.60
Sodium molybdate	$Na_2MoO_4.2H_2O$	50	0.25
Copper sulphate	$CuSO_4.5H_2O$	5	0.025
Cobaltous chloride	$CoCl_2.6H_2O$	5	0.025
Stock 3 (iron-EDTA)			
Ferrous sulphate	$FeSO_4.7H_2O$	5560	27.8
Ethylene diamine tetraacetic acid disodium salt (dihydrate)	$Na_2EDTA.2H_2O$	7460	37.8
Stock 4 (organics)			
Meso-inositol	$C_6H_{12}O_6$	20,000	100
Nicotinic acid (pyridine-3-carboxylic acid)	$C_6H_5NO_2$	100	0.5
Pyridoxine HCl (vit. B6)	$C_8H_{11}NO_3.HCl$	100	0.5
Thiamine HCl (vit.B1)	$N_{12}H_{17}ClN_4.5HCl$	100	0.1
Glycine	$C_2H_5NO_2$	400	2.0

Note: All the stocks were prepared in the double distilled water

reported using mature tree explants in *Eucalyptus impensa* (Bunn 2005), *Holarrhena antidysenterica* (Mallikarjuna and Rajendrudu 2007), *Tectona grandis* (Akram and Aftab 2008), etc.

Organogenesis involves the formation of adventitious and axillary shoots. The adventitious shoot production comprises de novo shoot meristem formation from either callus or directly from organized tissue such as epidermal or subepidermal cells. The axillary shoot production includes shoot formation from axillary buds, shoot tips and meristems (Jaiswal 2003). In organogenesis unipolar structure either shoots or root meristem is formed, whereas in somatic embryogenesis, bipolar structures are formed, possessing both shoot and root meristems (Litz and Gray 1992). Thus organogenesis is a multistage, labour-intensive and time-consuming process. As compared to organogenesis, somatic embryogenesis provides an ideal experimental process for investigation of plant differentiation (Litz and Gray 1992). Many workers have emphasized somatic embryogenesis as a preferred method for

Table 4.2 Composition of WPM (Lloyd and McCown 1980) medium

Chemicals name	Chemical formula	Amount in stock (mg/l)	Amount in medium (mg/l)
Stock 1 (macronutrients)			
Ammonium nitrate	NH_4NO_3	8000	400
Potassium sulphate	K_2SO_4	19,800	990
Calcium nitrate tetrahydrate	$Ca(NO_3)_2.4H_2O$	11,120	556
Calcium chloride dihydrate	$CaCl_2.2H_2O$	1920	96
Magnesium sulphate heptahydrate	$MgSO_4.7H_2O$	7400	370
Potassium dihydrogen phosphate	KH_2PO_4	3400	170
Stock 2 (micronutrients)			
Boric acid	H_3BO_3	1240	6.2
Manganese sulphate heptahydrate	$MnSO_4.7H_2O$	4460	22.30
Zinc sulphate heptahydrate	$ZnSO_4.7H_2O$	1720	8.60
Sodium molybdate dihydrate	$Na_2MoO_4.2H_2O$	500	0.25
Copper sulphate pentahydrate	$CuSO_4.5H_2O$	500	0.25
Stocks 3 (iron-EDTA)			
Sodium EDTA dihydrate	$Na_2EDTA.2H_2O$	7460	37.30
Ferrous sulphate heptahydrate	$FeSO_4.7H_2O$	5560	27.80
Stock 4 (organics)			
Meso-inositol	$C_6H_{12}O_6$	20,000	100
Glycine (amino acetic acid)	$C_2H_5NO_2$	400	2.0
Thiamine HCl (vit. B1)	$N_{12}H_{17}ClN_4.5HCl$	20	0.1
Nicotinic acid (pyridine-3-carboxylic acid)	$C_6H_5NO_2$	100	0.5

Note: All the chemicals weigh, dissolved separately in double distilled water, mixed together and maintained the final volume 1000 ml. All the stocks were prepared in the double distilled water

genetic improvement and multiplication of valuable germplasm of a number of woody perennials (Gupta and Durzan 1987; Bhansali et al. 1991) because somatic embryos have always been assumed to originate from a single cell and have a closed vascular system (Litz and Gray 1992).

4.4 Regeneration via Somatic Embryogenesis

Somatic embryogenesis is the process by which somatic cells, under inductive conditions, generate embryogenic cells, which undergo a series of morphological and biochemical changes ensuing the formation of somatic embryos (Zimmerman 1993; Komamine et al. 2005). Somatic embryogenesis is a developmental pathway that contains the several steps (Fig. 4.2) in which somatic cells form a bipolar structure which resembles with zygotic embryos, but fusion of gametes does not take place (Jimenez 2001).

Fig. 4.2 Schematic diagram of somatic embryogenesis (*SEs* somatic embryos)

It is the most effective method of micropropagation to produce true-to-true type plantlets of valuable and elite genotype for large-scale multiplication by producing a large number of somatic embryos from single explants in a short time period (Lema-Ruminska et al. 2013). When integrated with conventional breeding programmes and molecular and cell biological techniques, somatic embryogenesis provides a valuable tool to enhance the pace of genetic improvement of commercial crop species (Stasolla and Yeung 2003). Somatic embryogenesis is an ideal system for investigation of the whole process of differentiation from a single cell to plant, as well as of the mechanism of expression of totipotency in plant cells. It has the following major advantages as compared to zygotic embryogenesis:

1. The process of somatic embryogenesis can be easily monitored.
2. The environment and the development phase of somatic embryo can be controlled.
3. A high number of embryos can be easily achieved from the single explants (Komamine et al. 2005).

These characteristics have made somatic embryogenesis provide an excellent morphogenetic system for investigating the cellular and molecular process underlying differentiation (Benelli et al. 2001), and it is also a model system for studying the genetic basis of embryogenesis and totipotency of somatic cells (Kurczyjska et al. 2007). In addition, somatic embryogenesis also provides the possibility to produce artificial seeds and valuable tools for genetic engineering and germplasm conservation by cryopreservation (Litz and Gray 1995; Ara et al. 2000).

Somatic embryogenesis can be induced in two different pathways named direct somatic embryogenesis (DSE) and indirect somatic embryogenesis (ISE). DSE is when a minimal proliferation of unorganized tissue precedes the direct embryo formation from the surface of explant, while in ISE embryos are formed from the proliferated callus (Sharp et al. 1980). In DSE, proembryogenic competent cells are already present, and the expression of the embryogenetic programme simply depends on suitable conditions. Direct embryogenesis devised from pre-embryogenic determined cells (PEDCs), whereas indirect somatic embryogenesis originated from induced embryogenic determined cells (IEDCs), and callus involves an interphase before formation of embryos. Embryogenesis from PEDCs requires only a suitable in vitro environment to follow the requisite pattern of cell division (Thorpe 1988; Bhojwani and Razdan 1996). Since tissues of mature trees are highly differentiated, embryogenesis from such tissues must proceed via induced embryogenic determined cell "IEDC" (Sharp et al. 1980; Merkle 1995). These highly differentiated tissues must undergo major epigenetic changes to initiate somatic embryo production; therefore IEDC requires an in vitro environment initially for dedifferentiation and then redifferentiation of quiescent cells to an embryogenic state. Direct and indirect embryogenesis are two additional terms used to describe PEDC and IEDC embryogenesis, respectively. In DSE somatic embryos appear directly over the explants without the intermediate callus phase, so, very less time is required for the complete plantlets formation, whereas in ISE, a long intermediate callus phase is involved before the appearance of somatic embryos and required more time for plantlets formation (Michler and Bauer 1991; Rugini and Caricato 1995; Berthoulay and Michaux-Ferriere 1996). The callus formed may not be always embryogenic but serves as the precursor tissue for differentiation of embryogenic callus (Emons 1994).

A number of media have been used for the induction of embryogenic cultures. However, most of the successful reports are based on Murashige and Skoog's (1962) (MS) medium. The effect of medium composition on somatic embryogenesis has been demonstrated in citrus (Gill et al. 1995), papaya (Fitch and Manshardt 1990), mango (Ara et al. 2000), etc. Success of regeneration through somatic embryogenesis depends upon the several factors such as right choice of the explant, genotype, preconditioning, explant orientation and media requirements (Thorpe et al. 1991; Litz and Gray 1992) and also the endogenous level of hormones in the explants and application of appropriate plant growth regulators.

Hormones play a vigorous role in the regulation of developmental switches. Details of plant growth regulators, preparation and their respective solvent are mentioned in Table 4.3. Auxins and cytokinins are the key plant growth regulators

Table 4.3 Plant growth regulators used in the present investigation

Name of plant growth regulators		Solvent(s)
Auxins	2,4-Dichlorophenoxyacetic acid (2,4-D)	Absolute ethyl alcohol
	Indole-3-acetic acid (IBA)	Absolute ethyl alcohol
	Indole-3-butyric acid (IAA)	Absolute ethyl alcohol
Cytokinins	6-Benzylaminopurine (BAP)	0.1N NaOH
	Kinetin (Kn)	0.1N NaOH
	Zeatin (ZN)	0.1N NaOH
	Thidiazuron (TDZ)	0.1N NaOH
Others	Abscisic acid (ABA)	0.1N NaOH

involved in the regulation of cell division and differentiation. 2, 4-Dichlorophenoxyacetic acid (2, 4-D) is an auxin, frequently exogenously used for induction of somatic embryogenesis (Dudits et al. 1995; Von-Arnold et al. 2002). Although cytokinins have sometimes been incorporated into the induction medium, they are probably not critical for induction. The mechanism of somatic embryo induction from cells is still unclear. Auxins were shown to play a vital role in initiation and maintenance of cell division (Dudits et al. 1995) that ultimately leads to somatic embryo formation. Either lowering or complete removal of auxin is necessary for the continuous development and maturation of the somatic embryo (Nomura and Komamine 1995; Raghavan 2000). The synchronized activation of auxin and stress responses may be a key incident in cellular adaptation, causing genetic, metabolic and physiological reprogramming that develop the embryogenic competence (totipotency) of somatic plant cells (Feher et al. 2003).

The first report of somatic embryogenesis was documented *Oenanthe aquatica* (Waris1957) and *Daucus carota* (Reinert 1958; Steward 1958). In woody (angiosperms) plant, it was first time reported in *Santalum album* (Rao 1965). Although somatic embryogenesis has been reported for several tree species, the quality of somatic embryos with regard to their germination ability or conversion into plants has been very poor. Sometimes the normal-looking somatic embryos may have incomplete development. In tree species, germination of viable plantlets from somatic embryos is a major problem than development and maturation of somatic embryos. The problem may occur at any stage of development like maturation, germination, shoot apex elongation or acclimatization. In contrast to zygotic embryos, the somatic embryos usually do not go through the final phase of embryogenesis, i.e. embryo maturation.

Maturation is a crucial phase between embryo development and germination (Quatrano 1989). Embryo maturation is characterized by the accumulation of reserve food materials and proteins which deliver the desiccation tolerance to the embryos (Bhojwani and Dantu 2013). Usually abscisic acid (ABA) regulates the synthesis and deposition of storage and late embryogenesis abundant (LEA) proteins during somatic embryogenesis (Dodeman et al. 1997). Abscisic acid (ABA) promotes normal development of embryos by suppression of secondary embryogenesis and multiple cotyledons (Ammirato 1983). ABA also helps in accumulation of sugars and late embryogenesis abundant (LEA) proteins and finally promotes the acquisition of desiccation tolerance (Buitink and Leprince 2008). ABA promotes the transition from multiplication to maturation stage and inhibits precocious germination (Langhansova et al. 2004). Cytokinin can also be important for somatic embryo maturation and has been demonstrated to influence development of cotyledon and shoot apex (Litz and Gray 1992). The role of some factors such as amino acids (Lai and McKersie 1993; Khlifi and Tremblay 1995; Robichaud et al. 2004), sucrose (Linossier et al. 1997), PEG (Capuana and Debergh 1997; Langhansova et al. 2004) and ABA (Arnold and Hakman 1988; Linossier et al. 1997; Sagare et al. 2000) in the maturation of somatic embryos is well reported.

In tree species, somatic embryogenesis process also faces several limitations such as low-frequency somatic embryo production, abnormal embryo formation, incomplete and low-frequency maturation, low germination and less survival in field conditions (Merkle 1995). Although somatic embryogenesis has been reported in several tree species, the quality of somatic embryos with regard to their germination ability or conversion into plants needs to be improved. The important factors contributing to the success in regeneration through somatic embryogenesis are the right choice of the explant, genotype, preconditioning, explant orientation and media requirements, etc. (Thorpe et al. 1991; Litz and Gray 1992; Bhojwani and Dantu 2013). A number of successful reports of somatic embryogenesis in tree species such as *Prosopis laevigata* (Buendía-González et al. 2012), *Eucalyptus globulus* (Andrade et al. 2011), *Phellodendron amurense* (Azad et al. 2009), *S. trifoliatus* (Asthana et al. 2011), *Kalopanax pictus* (Moon et al. 2005), etc. have been reported.

4.5 Advantages and Limitations of Micropropagation Techniques

Propagation of woody trees through tissue culture provides many advantages over conventional propagation methods such as rapid multiplication of valuable genotypes, production of improved cultivars and disease-free plants, season-independent production and conservation and exchange of germplasm (Komamine et al. 2005; Asthana et al. 2011). SE also offers the opportunity to produce artificial seeds and appreciated tools for genetic engineering and germplasm conservation by cryopreservation (Litz and Gray 1995; Ara et al. 2000). Besides the several advantages, few limitations also exist with propagation of tree species through tissue culture such as phenolic exudation, contamination of different levels and seasonal dependency for collection of explants.

4.6 Propagation of *Sapindus* Species

4.6.1 Conventional Propagation of *Sapindus*

Conventionally, *Sapindus* species is grown by either seed germination or stem cuttings. Propagation rate was very less through conventional stem cutting methods. Seed germination rate of *Sapindus* seed is very less due to presence of hard seed coat, and germination process is very slow. The seeds have hard seed coats due to which they become dormant (Singh et al. 2010). Propagation through seed is unreliable due to the much reduced survival of seedling because of high mortality rate at seedling stage affected by the environmental factors (Philomina 2010).

Seed germination behaviour of *S. mukorossi* was also studied by Dobhal et al. (2012). They used boiling water, cow dung slurry and concentrated HCl as pretreatment agents for seed before the germination for different time durations. These pretreatments significantly affected the percentage of seed germination. The highest seed germination percentage (66.6%) was recorded from the pretreated seeds with concentrated HCl for 110 min. followed by boiling water for 5 min. + cold water for 48 h treatment (43.33%) and cow dung slurry for 2 days (20.00%). Seed germination percentage was very less (6.66%) in control when compared to the experimental data.

4.6.2 Micropropagation Studies

Organogenesis and somatic embryogenesis are the two most important pathways of in vitro cultures. Besides these two pathways, single cell culture, suspension culture and hairy root culture can be opted for the regeneration of in vitro plants. A number of workers established the effective regeneration protocol of *Sapindus species* using the various explants through either organogenesis or somatic embryogenesis. In brief the micropropagation studies, viz. organogenesis and somatic embryogenesis, are demonstrated in Tables 4.4 and 4.5, respectively.

4.7 Mass Propagation of *Sapindus mukorossi*

4.7.1 Organogenesis

First report of organogenesis was made by Philomina and Rao (2000). Seedlings were raised aseptically by inoculating seeds on agar-water medium. They took apical and axillary meristems from in vitro raised seedlings as explants and successfully induced organogenesis. Maximum shoot formation in both explants was observed on MS medium supplemented with BAP (0.4 μM) and GA_3 (2.8 μM), and rooting of shoots was observed on MS medium containing IBA. The seedling-derived explants are considered as unproven materials, and eliteness of plant can be checked only after occurrence of flowering and fruit setting. Nodal explants from field grown

Table 4.4 Details of organogenesis of *Sapindus species*

Plant name	Explant	Explant source	Frequency/ no. of shoots	Media for shoot induction	Rooting (%)	Rooting media	Survival (%)	References
S. mukorossi	Apical and axillary meristem	1-month-old sapling	5–8	MS + BAP (0.4 μM) + GA_3 (2.8 μM)	68	MS + IBA (3.4 μM)	60	Philomina and Rao (2000)
	Node	Field grown plants	2–3	MS medium having BAP (2.0 mg/l) + NAA (1.0 mg/l)		MS + 2 mg/l		Singh et al. (2010)
S. trifoliatus	Node			MS + BAP (4.44 μM)		LMS + 1 mg IBA (for 24 h) to MS basal medium	90	Asthana et al. (2011)

Table 4.5 Details of somatic embryogenic pathways of *Sapindus species*

Plant name	Explant used	Optimum growth regulator used				Plantlets survival (%)	References
		Callusing (%)	Embryo development (No.)	Embryo maturation (%)	Embryo germination (%)		
S. mukorossi	Leaf	MS + BAP (8.88 μM)	(MS + BAP (8.88 μM)	(MS + BAP (8.88 μM)	MS + BAP (8.88 μM)		Singh et al. (2015)
	Leaf and petiole	NA (MS + different ranges of Kn and KNO_3)	61.66 ± 9.0 (MS + 1500 mgl^{-1} KNO_3 + 0.3 mgl^{-1}Kn)	15.00 ± 2.16 (MS + 0.3 mgl^{-1} GA3 + 1000 mgl^{-1} CH)	NA (MS + 100 mgl^{-1} NH_4Cl + 2,0 mgl^{-1} Kn)	NA	Dobhal et al. (2012)
	Leaf (from 60 days old aseptically grown seedlings)	82.5 (MS + 2,4-D (6.7 μM) + Kn (9.0 μM)	40–50		90 (MS basal)	90	Philomina (2010)
	Leaf (from in vitro seedlings)	53.9 ± 4.4 (B5 + 0.1 mgl^{-1} 2,4-D + 0.01 mgl^{-1} BAP)		73.3 ± 8.1 (B5 + ABA 5 mgl^{-1})	90	Not mentioned	Kim et al. (2012)
	Rachis	MS media + BAP (8.88 μM)	MS media + BAP (8.88 μM)	MS media + ABA (0.0378 μM)	MS media + BAP (8.88 μM)	80	Singh et al. (2016a, b, c)
	In vitro leaf	MS media + BAP (8.88 μM)	MS media + BAP (8.88 μM)	MS media + 4% sucrose + ABA (0.0378 μM)	MS medium +sucrose (4%) + BAP (4.44 μM)	80	Singh et al.
S. trifoliatus	Sepals	2, 4-D (5 mg/l) + KN (0.1 mg/l)	MS media + 200 mg/l	MS media	MS media (2% sucrose)		Asthana et al. (2017)

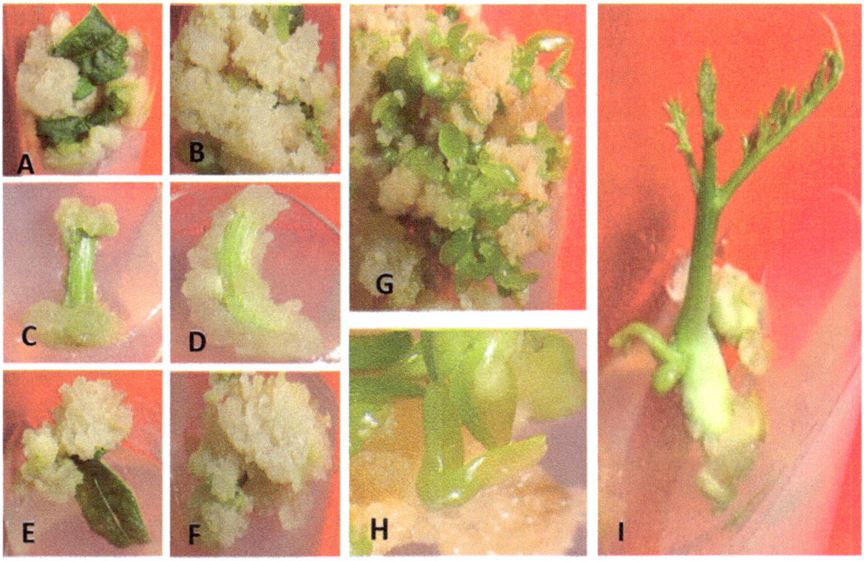

Fig. 4.3 Somatic embryogenesis from different parts of *S. mukorossi*. (**a**) In vivo leaf, (**b**) callus proliferation from in vivo leaf, (**c**) rachis, (**d**) callus proliferation from rachis, (**e**) in vitro leaf, (**f**) callus proliferation from in vitro leaf explant, (**g**) somatic embryo development, (**h**) cotyledonary-stage embryos and (**i**) germination of somatic embryos

plants were used for indirect organogenesis by Singh et al. (2010), but age of plants was not mentioned in the publication. They used MS media supplemented with different concentrations and combinations of 2, 4-D and BA for callus initiation. The maximum number of shoots (2.6 ± 0.25) and shoot length (2.2 ± 0.25) was achieved on MS medium having BAP (2.0 mg/l) + NAA (1.0 mg/l). The highest number of roots was observed on half strength media supplemented with 2 mg/l IBA after 22 days. They reported 60% survival frequency under field condition.

4.7.2 Somatic Embryogenesis

In *S. mukorossi* various plant parts such as in vivo leaf, in vitro leaf, seedling-derived leaf, petiole and rachis have been used for the development of somatic embryogenesis system. Somatic embryogenesis pattern was described in Figs. 4.3 and 4.4.

4.7.2.1 Seedling-Derived Leaf Explants

The first and preliminary report of somatic embryogenesis was made by Philomina (2010). He used leaves of in vitro raised seedlings as explants. Leaves inoculated on MS medium with 2,4- D (2.2 μM) and Kn (4.6 μM) showed embryogenic response within 1 week of inoculation. These embryos after maturation germinated into normal plantlets. The publication lacked information on maturation, germination and the optimization of different factors. Kim et al. (2012) also used leaf explants

Fig. 4.4 Hardening and field transfer of in vitro raised plants. (**a**) Complete plantlet, (**b** and **c**) hardening in culture condition, (**d**) acclimatization and (**e**) field transfer of plants

taken from 5 to 14 days old in vitro raised seedlings. They observed B_5 medium (Gamborg's medium) supplemented with 2,4-D (0.1mgl^{-1}) and BA (0.01mgl^{-1}) as most responsive for the induction of embryogenic callus. ABA (5mgl^{-1}) was found to be a promoter for the maturation of somatic embryos (Kim et al. 2012).

Similarly, Dobhal et al. (2012) observed somatic embryogenesis in leaf and petiole explants of *S. mukorossi*, but details about donor plant were not mentioned in their publication. They observed MS medium supplemented with Kn (0.3 mg/l) and KNO_3 (1500 mgl^{-1}) as the most responsive induction medium. For maturation, the medium was supplemented with GA_3 and casein hydrolysate (CH). The explants (other than zygotic embryos) taken from mature trees are mostly used for the regeneration of elite plants, as they are considered proven material and less variation is observed in such protocols.

4.7.2.2 In Vivo Leaf Explants

In vivo leaf explant was used for development of somatic embryogenesis system in *S. mukorossi* (Singh et al. 2015). Fresh flushes of leaves were collected from a mature tree, washed and sterilized and used as explants for establishing a reliable and large-scale multiplication protocol. Leaf explants were inoculated on the MS medium supplemented with different plant growth regulators. MS medium

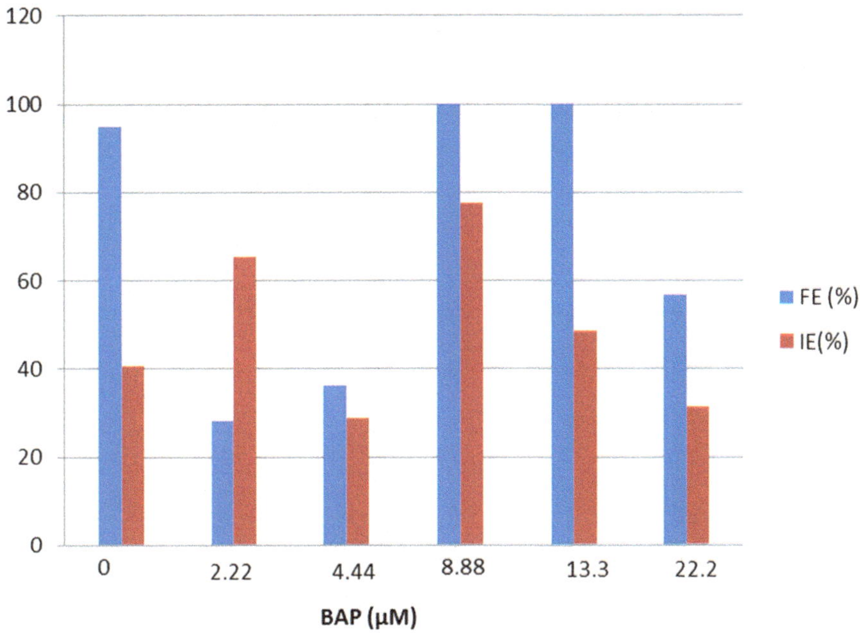

Fig. 4.5 Development of somatic embryos from in vivo leaf explants (*IE* intensity of embryogenesis, *FE* frequency of embryogenesis)

containing different concentrations of 2,4-D acid or BAP was able to induce the callusing from the midrib region of leaf explants. Size, physiological age, orientation of leaf explants on the culture medium and concentrations and type of plant growth regulators significantly affected the callus induction and somatic embryogenesis. Adaxial-side-up orientation of leaf explants significantly promoted embryogenesis when compared to the abaxial-side-up orientation. A similar result of orientation of explant was documented in *Peucedanum oreoselinum* (Coste et al. 2011). Adaxial-side-up treatment may be endorsed to the natural facility of conduction of nutritional and growth factors from the adaxial to the abaxial surface (Corredoira et al. 2012). Leaf explants with midrib portion have the high efficiency to induce callusing because midrib region have highly dividing cells (Asthana et al. 2011), whereas those explants that lack the midrib were unable to induce the callus at either side of the leaf surface. In the present study, the medium with 2,4-D was able to induce the callusing, but after the subculture, these calli are either unable to induce the formation of somatic embryos or produced very less or poor somatic embryos. The highest frequency of somatic embryogenesis (100%) and average maximum number of somatic embryos (77.36) were observed on MS medium supplemented with 8.88 μM BAP (Fig. 4.5). The highest frequency of maturation (85.1%) and highest germination of somatic embryos (87.25) were recorded on MS medium containing 8.88 μM BAP than on medium devoid the BAP (Fig. 4.6).

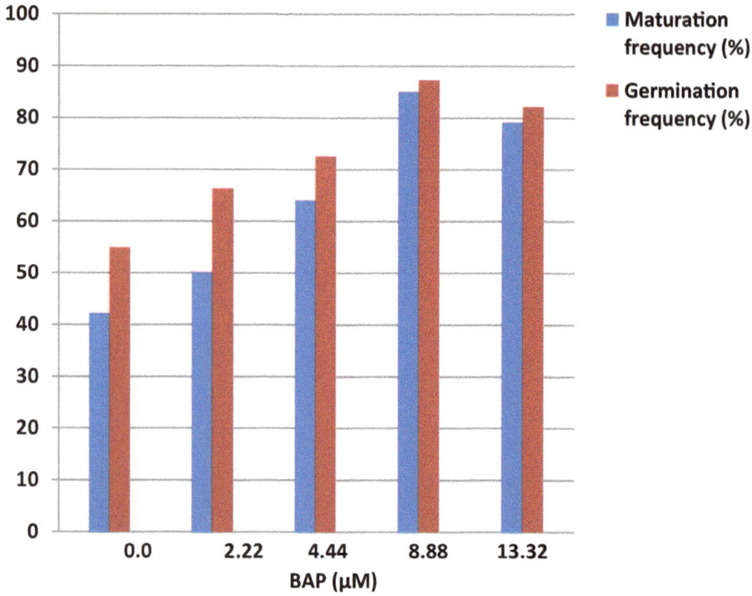

Fig. 4.6 Maturation and germination of somatic embryos developed from in vivo leaf explants

Maturation and the efficiency of germination are major factors for the success of plant regeneration (Martinelli and Mandolino 2009; Bhojwani and Dantu 2013). Of the germinated embryos, about 74.26% formed the complete plantlets. Sagare et al. (2000) also reported that MS medium supplemented with different cytokinins (BAP, KIN and ZN) was the best media for callus induction, somatic embryogenesis and plant regeneration in *Corydalis yanhusuo*. MS medium with supplementation of 8.88 μM BAP was the best media for callus induction, somatic embryo development, maturation, germination and plantlet conversion from leaf explants of *S. mukorossi*. About 90% plantlets survived after successful acclimatization and transplantation in the field (Fig. 4.4).

All the stages of somatic embryos and embryogenic and nonembryogenic calli were subjected to the scanning electron microscopic (SEM) analysis. SEM photographs showing the origin of somatic embryos from embryogenic callus and the development of globular-,

heart- and cotyledonary-stage somatic embryos (Singh et al. 2015). SEM analysis helps to investigate the ontogeny, structure and differentiation of somatic embryos and also contributed to the understanding of basic mechanisms involved in the achievement of embryogenic capability (Rocha et al. 2012; Aslam et al. 2014).

4.7.2.3 Rachis Explants

In plants a rachis is the main axis of a compound structure. It can be the main stem of a compound leaf such as acacia or fern. Rachis is an intersegment of petiole axis between two successive leaflets in a compound leaf (Fig. 2.1, Chap. 2). Very few

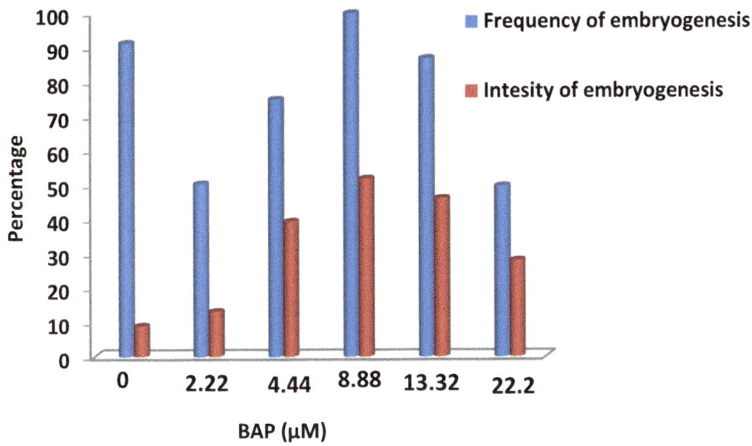

Fig. 4.7 Development of somatic embryos from rachis explant

reports are available of the micropropagation from the rachis explant. There is infinite opportunity to explore the rachis for mass multiplication using the various plant tissue culture techniques.

The present study was performed using the rachis as explant and successfully established the efficient somatic embryogenesis protocol for regeneration of the number of plant in a short time period (Singh et al. 2016a, b, c). Rachis was collected from newly growing leaves of a mature tree and sterilized using the standard protocol of sterilization. Sterilized rachis was inoculated on the MS medium supplemented with different types and concentrations of plant auxins and cytokinins. Callus initiation was observed on MS medium supplemented with TDZ (1-Phenyl-3-(1, 2, 3-thiadiazol-5-yl) urea), zeatin or BAP. These calli transferred on MS basal medium and MS medium + 8.88 μM BAP for induction of somatic embryogenesis.

Both the media have the ability to induce the somatic embryogenesis, but MS medium + 8.88 μM BAP have more potential in comparison to the MS basal medium.

Frequency of embryogenesis and intensity of embryogenesis were observed 100% and about 52%, respectively, on MS medium + 8.88 μM BAP (Fig. 4.7). After the development of different stages of somatic embryos, early cotyledonary embryos were transferred on MS medium + 8.88 μM BAP with different concentrations of ABA for 21 days. Among all the ABA concentrations, 0.0378 μM ABA favoured the maximum maturation of somatic embryos about 93% (Fig. 4.8). MS medium supplemented with 8.88 μM BAP also promotes the maximum germination of somatic embryos (77%) (Fig. 4.9). Germinated embryos converted into complete plantlets. As earlier mentioned the in vitro raised plantlets were hardened, acclimatized and transferred to the field with a survival frequency of 80%.These plantlets were morphologically similar to the mother plants.

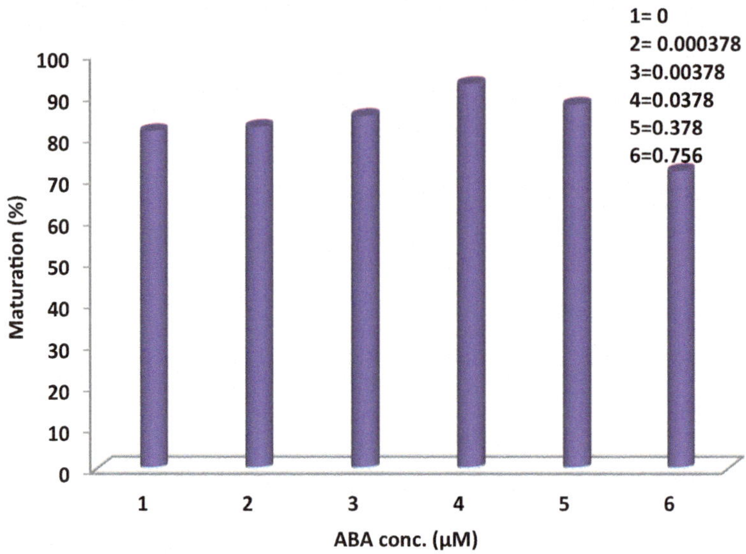

Fig. 4.8 Maturation of somatic embryos developed from rachis explant

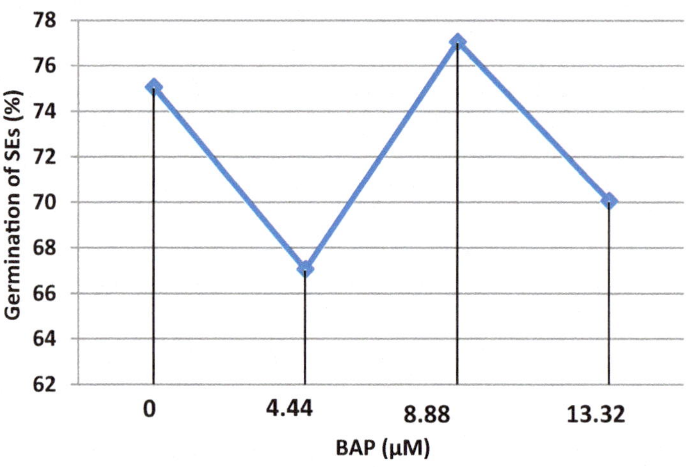

Fig. 4.9 Germination of somatic embryos developed from rachis explant

4.7.2.4 In Vitro Leaf Explants

In vitro leaf of *S. mukorossi* plantlets was also used as explant to establish the somatic embryogenic system (Singh 2015), and a successful in vitro regeneration protocol has been developed for mass multiplication of this elite tree species. In vitro leaves were isolated from the micropropagated plantlets that rose from in vivo leaf explants. This experiment was performed to investigate the somatic embryogenic potential of in vitro leaf.

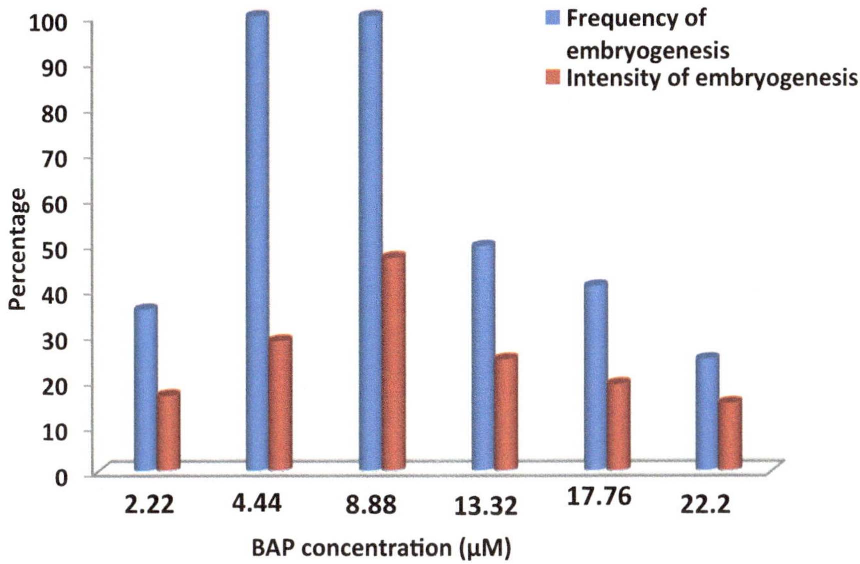

Fig. 4.10 Development of somatic embryos from in vitro leaf explant

On average 100% explants induced the callusing on MS medium supplemented with 8.88 µM BAP. Calli were transferred on MS basal medium for development of somatic embryos, and the average number of somatic embryos was 47.10/ subcultured callus (Fig. 4.10). A higher level of sucrose and supplementation of ABA promote the somatic embryo maturation. A maximum maturation percentage of somatic embryos was found on MS medium supplemented with 4% sucrose and 0.0378 µM ABA (Fig. 4.11). MS basal medium supplemented with 4% sucrose and 4.44 µM BAP promotes the germination of somatic embryos, and maximum germination percentage was recorded about 77.79% of matured embryos and converted into the complete plantlets (Fig. 4.12). Lower concentration of BAP favoured the germination of somatic embryos in comparison to the callus induction medium (MS medium + 8.88 µM BAP); it may be due to the presence of high concentration of endogenous cytokinins in the embryos. Efficacy of higher level of sucrose also reported in *Psidium guajava*, and application of BAP for the germination of somatic embryos was also reported in *Quercus rubra* (Vengadesan and Pijut 2009). As earlier mentioned the plantlets were transferred in field condition after the successful hardening and acclimatization. The field transferred plants grew well and morphologically similar with mother plants.

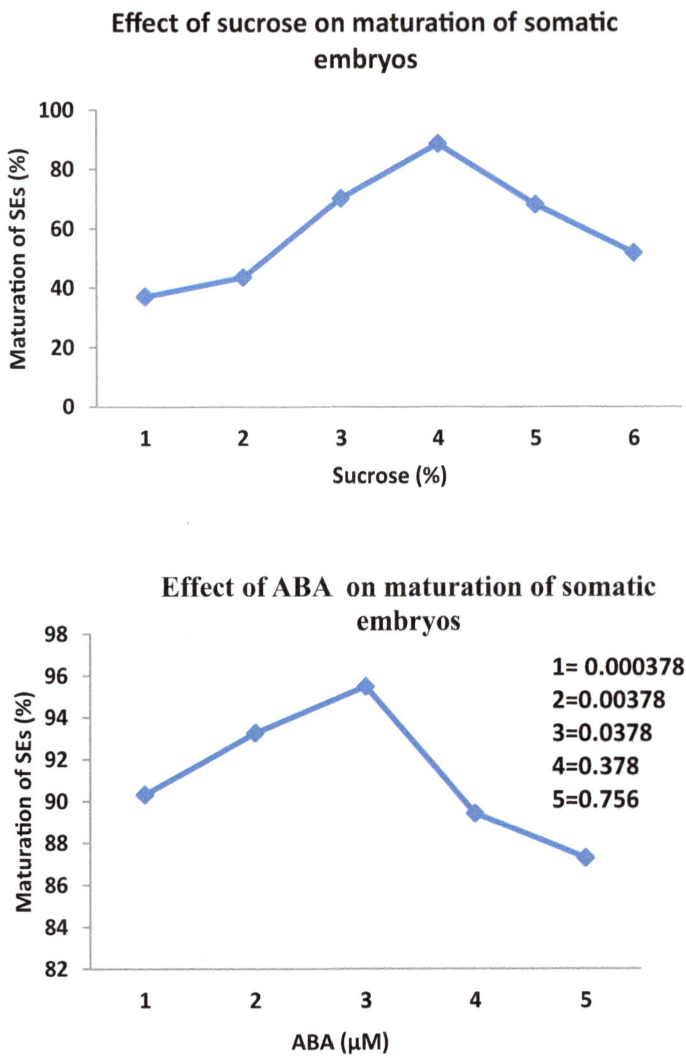

Fig. 4.11 Maturation of somatic embryos regenerated from leaf explant

4.8 In Vitro Regeneration in *Sapindus trifoliatus*

In vitro regeneration of *S. trifoliatus* has been done through both the methods, i.e. organogenesis and somatic embryogenesis. In coming paragraphs these are described in details.

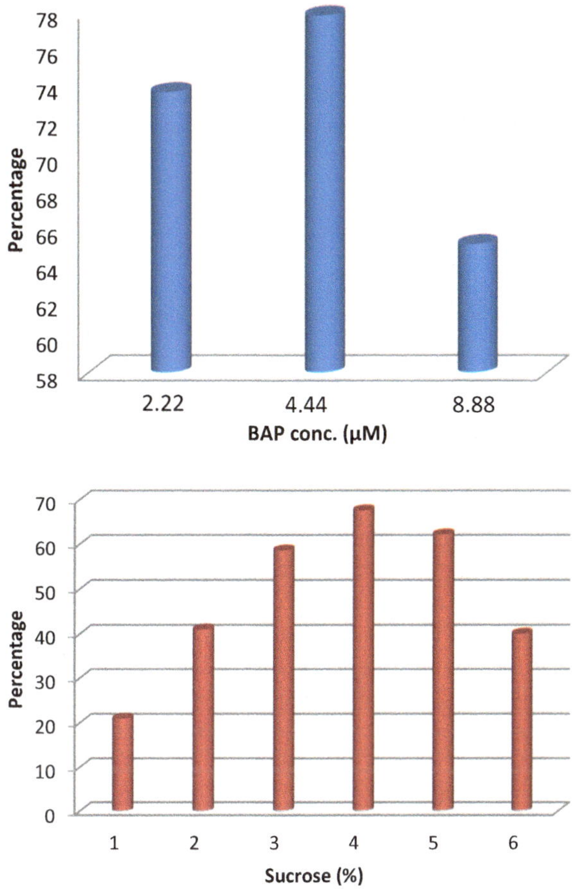

Fig. 4.12 Germination of somatic embryos developed from in vitro leaf explant

4.8.1 Organogenesis

Organogenesis is a primary pathway for the in vitro regeneration. Organogenesis pathway is less complex and easy in handling and required less time and labour when compared to the somatic embryogenesis. A successful in vitro regeneration protocol has been established for rapid micropropagation using nodal segments of seedlings (Asthana et al. 2011). Nodal explants harvested from 4-week-old seedlings were most responsive for shoot induction. Induction of shoots was observed after the 10–15 days of inoculation. Maximum shoot induction (97.22%) was observed in MS media supplemented with 1.0 mg/l BAP while lower response of shoot induction was observed in kinetin, zeatin and lower concentrations of BAP. Newly formed shoots were collected and the mother explants were subcultured again to the same medium subsequently which produced 5.16 shoots/explants after the third subculture. In vitro nodal explants (excised from in vitro raised plantlets) produced the 6.89 shoots/explants transferred on the same medium after the third subculture. In vitro grown shoots showed the highest rooting (91.67%) when it was treated with liquid MS medium with 1.0 mg/l IBA for 24 h and then transferred to the agar-solidified MS basal medium. The in vitro raised plantlets were successfully acclimatized and transplanted to the field condition, and 90% survival rate was recorded. Result showed that shoot regeneration from seedling node explant was clearly affected by the seedlings' age, concentrations of growth regulators and successive transfer of the mother explant.

Selection of seedling as mother plant was also reported in some woody trees named *Terminalia chebula* (Shyamkumar et al. 2003), *Boswellia ovalifoliolata* (Chandrasekhar et al. 2005), *Pterocarpus santalinus* (Rajeswari and Paliwal 2006) and *Stereospermum personatum* (Shukla et al. 2009) plant regeneration.

4.8.2 Somatic Embryogenesis

4.8.2.1 Leaf Explant

The first report of somatic embryogenesis in *S. trifoliatus* was documented by Desai et al. (1986). At that time the plant tissue culture in their early stage and very few reports were available. Mature tree of *S. trifoliatus* L. (about 60 years old) was selected as mother plant. Young leaves were collected and used as explant for induction of somatic embryogenesis. Murashige and Skoog's medium supplemented with kinetin (0.5 mg/l) and 2,4-dichlorophenoxyacetic acid (2,4-D, 2 mg/l) was for callus induction. White friable callus was formed after the 10 days of inoculation of the explants. Lowering down the concentrations of 2,4-D during subsequent subcultures leads to the formation of embryoids. Embryoids were observed on medium containing the 0.5 mg/l 2,4-D and 0.5 mg/l kinetin. At the same time 2, 4-D induced the callusing in heart-shaped embryos; therefore embryos do not get to mature up to cotyledonary stage. Therefore, the entire callus bearing somatic embryos was first transferred to a medium supplemented with 2 mg/1 BAP and 0.5 mg/l kinetin. Somatic embryos were unable to germinate on MS basal media or MS media with cytokinins. Another supplement 5-methyl-tryptophan (tryptophan analogue) showed positive effect on germination of somatic embryos and prevention of re-callusing. 5-Methyl-tryptophan further enhanced the percentage of germination of somatic embryos when it was added in liquid media. About 400 plantlets were produced from a single leaf disc explant (5 mm diameter) during 19 weeks of subculturing (Desai et al. 1986).

4.8.2.2 Development of Somatic Embryos in Saline Media

Salinity effect on the development of somatic embryos (from leaf explant) was evaluated by adding the NaCl in the medium. Growth and development of somatic embryos were enhanced by adding 25 and 50 mol m^3 NaCI into the medium, and these concentrations favoured the secondary embryoid production, while addition of 100 mol m^3 NaCI to the medium did not favour the growth and development of somatic embryos. Increasing the concentration of NaCl (200 mol m^3) in the medium causes the complete death of proembryo structures within 7 days of subculturing.

Analysis of somatic embryos revealed that upon salinization, they accumulated Na^+ and Cl^-in significant amounts, but the content of Na^+ was much less compared to that of Cl^-. In somatic embryos K^+ content increased after addition of NaCI (up to 50 mol m^3) in the medium. In response to the saline stress, content of proline in somatic embryos was increased. The quantity of free sterols, steryl glycosides, steryl esters and phospholipids also increased in salt-affected somatic embryos. The results suggest that somatic embryos of *S. trifoliatus* can tolerate concentrations of NaCI up to 100 mol m^3 without affecting growth and that they have sufficient cellular mechanisms to tolerate salinity at relatively high levels.

4.8.2.3 Sepal Explant

Induction of somatic embryogenesis and regeneration of in vitro plants were established in using sepal explants through callus culture techniques (Asthana et al. 2017). Explants inoculated on MS medium containing 5.0 mg/l, 2,4-D and 0.1 mg/l kinetin produced embryogenic calli. Whitish or greenish nodular embryogenic structure was formed from the embryogenic calli on the MS basal medium containing 200 mg/l L-glutamine. These nodular embryogenic structures finally contributed in the formation of somatic embryos or secondary nodular embryogenic structures or both. Nodular embryogenic structures separated from embryogenic calli and transferred to MS medium containing 200 mg/l L-glutamine produced maximum number of globular-, heart- and cotyledonary-stage somatic embryos. About 88.8% somatic embryos germinated and developed into plantlets on agar-solidified MS medium having 2% sucrose. Histological studies have exposed the development of somatic embryos (different stages) from nodular embryogenic structure. The embryogenic competence of the culture could be maintained by recurrent production of nodular embryogenic structures and somatic embryos. Orientation of explants played a significant role for the establishment of somatic embryogenetic system. About 90% plants were successfully hardened, acclimatized and survived in field condition. All the transferred plants develop new leaves and showed healthy growth. The secondary nodular embryogenic structure also inflated and contributed in the formation of somatic embryos and sometimes converted into the tertiary nodular embryogenic structure. The embryogenic potential of the cultures has been maintained for about 12 months via enduring production of nodular embryogenic structure and somatic embryos. Nodular embryogenic structures are constituted by tissues having high proliferative potential; thus embryogenic competence of explants can be sustained by continued proliferation of nodular proembryogenic structures as a form of repetitive embryogenesis (Martínez et al. 2015).

4.9 Genetic Fidelity Appraisal

Somaclonal variation can be defined as genetic variation perceived among progeny of plants regenerated from somatic cells cultured in vitro (Larkin and Scowcroft 1981). Although theoretically all plants regenerated from somatic cells should be clones and genetically similar, a number of observations have indicated that variants may be present among the progenies. Somaclonal variations mostly initiated due to stress caused by changes in culture conditions and final changes occur in DNA in the form of point mutations, chromosome rearrangements, DNA methylation, etc. (Phillips et al. 1994; Martin et al. 2006). Generally in callus culture, somaclonal variation can occur due to loss of some desirable characters. Most of this variation may be related to substances in the culture media that act as mutagens or maintain a degree of cell division for which the plant cell's repair mechanisms cannot work properly. Growth regulators may also be the ancillary basis for quantitative and structural chromosome change. The use of plant growth regulators (PGRs), especially synthetic PGRs at sub- and super-optimal levels, may cause the somaclonal variations (Martins et al. 2004). The extent of genetic variation induced by micropropagation methods depends upon the source of the explants or the regenerants (i.e. cells, callus or protoplasts) and the method of regeneration selected (Larkin et al. 1989).

Maintenance of genetic integrity is a major concern for large-scale application of plant tissue culture techniques. Maintenance of clonal fidelity means no change in genomic constituents in a particular clone throughout its life span (Lattoo et al. 2006). However, the large-scale implication of any micropropagation methodology carries the risk of somaclonal variation among the subclones of one parental line (Larkin and Scowcroft 1981). The confirmation of the genetic fidelity among raw materials utilized in the preparation of medical formulations in the drug development programme is essential to maintain the consumer's confidence. Thus to identify the somaclonal variations in the regenerants is essential at an early stage. Several techniques have been used for identification of somaclonal variants in in vitro raised plantlets. The detection of somaclonal variants among the micropropagated population can be done using morphological traits, physiological behaviour, karyotype analysis or isozyme profiles, but these techniques are not 100% reliable and have several limitations (Rani and Raina 2000). A limited number of informative makers and its proneness to developmental and environmental variations are the major constraints of isozyme technique (Rani et al. 1995). On the other hand, DNA-based molecular markers are more informative and can easily detect somaclonal variations that may be present in in vitro plantlets (Varshney et al. 2001), because molecular markers are not altered during environmental changes. So, they have been used in breeding programmes. Somaclonal variation can be easily detected by DNA fingerprinting using different types of molecular markers such as random amplified polymorphic DNA (RAPD), amplified fragment length polymorphism (AFLP), simple sequence repeats (SSR), intersimple sequence repeats (ISSR) or restriction fragment length polymorphism (RFLP). Among the various DNA-based markers, AFLP and RFLP are among the most reliable markers for screening of genetic fidelity in tissue-cultured plants. However, these methods involve the use of radioactive labelling, expensive enzymes and extensive care and, hence, may be unsuitable under certain conditions. Random amplified polymorphic DNA (RAPD) and intersimple sequence repeat (ISSR) analysis markers are very commonly used in crop and medicinal plants to analyse genetic stability in

Table 4.6 Genetic fidelity testing in different *Sapindus* species

Plant name	Method of regeneration	Explant used for regeneration	Source of explant	Molecular markers name	References
S. mukorossi	Somatic embryogenesis	Rachis	8–10-year-old tree	RAPD and ISSR	Singh et al. (2016a, b, c)
	Somatic embryogenesis	Leaf	8–10-year-old tree	RAPD and ISSR	Singh (2015)
	Somatic embryogenesis	In vitro leaf	18-month-old in vitro plant	RAPD and ISSR	Singh (2015)
S. trifoliatus	Organogenesis	Node	4-week-old seedling	RAPD	Asthana et al. (2011)

tissue-cultured plants as they are technically simple and quick to perform, require small amounts of DNA and have no requirement of prior information about the genome (Chuang et al. 2009; Sarmast et al. 2012). RAPD markers are the simplest technique to use for analyses of genetic fidelity among in vitro-propagated plants (Chuang et al. 2009; Sarmast et al. 2012). Polymorphism in RAPD markers results from a single nucleotide change that alters the primer binding site or an insertion or deletion within the amplified region. Such polymorphism usually results in the presence or absence of a distinct RAPD amplification product at a single locus (Tingey and Del-Tufo 1993). ISSR proved to be more effective and reproducible for detecting genetic uniformity (Martins et al. 2004), whereas RAPD marker is easy in handling and cost-effective also; therefore, most of the workers are using RAPD and ISSR in their studies. The molecular markers of genetic fidelity of *Sapindus* species are summarized in Table 4.6.

4.9.1 Genetic Fidelity in *Sapindus mukorossi*

4.9.1.1 In Vitro Raised Plantlets from Leaf Explants

The use of DNA-based markers allows efficient comparisons because genetic differences are detectable at all stages of development of the organism unlike allozymes which may show age-dependent changes. Detection of somaclonal variation at early stage is very important to avoid the economic loss (Chung et al. 2009) because commercial value of plant decreases due to somaclonal variation (Oropeza et al. 1995). Clonal fidelity was assessed using PCR-based RAPD and ISSR analysis.

A total of 16 primers (RAPD) were used for the production of scorable and reproducible bands. Amplified DNA profile of RAPD has shown monomorphic banding pattern of genomic DNA of mother and micropropagated plants. All primers generated a clear, scorable and reproducible banding pattern ranging from 350 to

Table 4.7 The nucleotide sequences of primers used for RAPD analysis

Primer	Sequence 5′-3′	No. of bands	Primer	Sequence 5′-3′	No. of bands
OPK-02	GTCTCCGCAA	4	OPL-01	GGCATGACCT	6
OPK-03	CCAGCTTAGG	9	OPL-04	GACTGCACAC	7
OPK-05	TCTGTCGAGG	6	OPL-20	TGGTGGACCA	2
OPN-01	CTCACGTTGG	6	OPO-06	CCACGGGAAG	6
OPN-16	AAGCGACCTG	5	OPP-06	GTGGGCTGAC	3
OPB-10	CTGCTGGGAC	4	OPZ-11	CTCAGTCGCA	4
OPB-12	CCTTGACGCA	3	OPZ-19	GTGCGAGCAA	6
OPB-18	CCACAGCAGT	8	OPV-07	GAAGCCAGCC	5

1250 bp in RAPD. All these RAPD primers produced a total of 84 bands, and a number of scorable bands for each primer varied from 2 (OPL-20) to 9 (OPK-03) with an average of 5.25 bands per primer (Table 4.7).

For ISSR molecular marker, 16 primers were used for generation of DNA profile. The DNA profile of ISSR showed monomorphic amplified bands of genomic DNA of both mother and regenerants. In ISSR primers the total number of scorable bands was 61 and varied for each primer from 2 (UBC-814, 842) to 6 (UBC-836, 848) with an average 3.81 bands per primer (Table 4.8). All primers generated clear, scorable and reproducible bands ranging from 250 to 1250 base pairs.

Successful implication of RAPD and ISSR for clonal fidelity testing has been reported in many micropropagated woody plants, viz. *Ochreinauclea missionis* (Chandrika and Rai 2009), *Spondias mangifera* (Tripathi et al. 2012), *Tylophora indica* (Sharma et al. 2014), *Terminalia arjuna* (Gupta et al. 2014), etc. RAPD and ISSR molecular markers have several advantages over various morphological, cytological and protein markers that are used for somaclonal identification in in vitro plantlets due to their simplicity, sensitivity, stability, cost-effectiveness, highly reproducible and reliability (Zietkiewicz et al. 1994; Ray et al. 2006). RAPD has proven to be quite efficient in detecting genetic fidelity even in closely related organism such as near-isogenic lines; ISSR technique is also a simple, fast, cost-effective and a highly reliable technique (Reddy et al. 2002) in which a single simple sequence repeat motif is used as a primer for amplification of regions between adjacent but palindromic microsatellites (Zietkiewicz et al. 1994). ISSR analysis involves the use of single sequence repeat motifs in order to prime the polymerase chain reaction and thereby amplify regions between adjacent, but inversely orientated, microsatellites. The microsatellite loci have been isolated. A (GA)n and (GT)n microsatellite-enriched library were developed to improve the type of molecular markers for genetic studies (Risterucci et al. 2005).

Table 4.8 The nucleotide sequences of primers used for ISSR analysis

Primer	Sequence 5′-3′	No. of scorable bands
UBC-802	ATATATATATATATATG	4
UBC-807	AGAGAGAGAGAGAGAGT	5
UBC-810	GAGAGAGAGAGAGAGAT	3
UBC-814	CTCTCTCTCTCTCTCTA	2
UBC-820	GTGTGTGTGTGTGTGTC	3
UBC-824	TCTCTCTCTCTCTCTCG	4
UBC-834	AGAGAGAGAGAGAGAGYT	2
UBC-836	AGAGAGAGAGAGAGAGYA	6
UBC-842	GAGAGAGAGAGAGAGAYG	2
UBC-848	CACACACACACACACARG	6
UBC-853	TCTCTCTCTCTCTCTCRT	5
UBC-862	AGCAGCAGCAGCAGCAGC	4
UBC-876	GATAGATAGACAGACA	3
UBC-887	DVDTCTCTCTCTCTCTC	4
UBC-889	DBDACACACACACACAC	5
UBC-899	CATGGTGTTGGTCATTGTTCCA	3

4.9.1.2 In Vitro Raised Plantlets from Rachis Explants

A total of 16 primers (10 ISSR and 6 RAPD) were used for the production of scorable and reproducible bands. RAPD and ISSR photographs are showing monomorphic amplified bands of genomic DNA of mother and micropropagated plants. Banding pattern was clear, scorable and reproducible ranges from 350 to 1250 bp in RAPD (Table 4.9) and 250 to 1250 bp in ISSR (Table 4.10). A total of 25 bands were produced from RAPD, a number of scorable bands for each primer varied from 2 to 8, and an average no. of bands was recorded 4.16 bands/primer. ISSR primers produced 28 scorable bands. A number of scorable bands varied for each primer ranged 1–4 with an average 2.8 bands per primer. RAPD and ISSR markers were most frequently used for assessment of clonal fidelity. RAPD marker assay is technically simple and rapid and requires small amount of DNA and due to these unique features has been preferred among other molecular markers (Ceasar et al. 2010). RAPD and ISSR gel photographs have shown monomorphic and clear DNA bands. RAPD and ISSR markers frequently have been used to appraise the genetic fidelity in a number of plants, viz. *Eclipta alba* (Ray and Bhattacharya 2010).

4.9.1.3 In Vitro Raised Plantlets from Leaf Explants

RAPD and ISSR analysis of in vitro raised plants showed similar banding pattern with mother plants that indicated the genetic integrity between them. RAPD and ISSR photographs showed clear, scorable and monomorphic bands. In RAPD, a total of 27 bands were observed with 6 primers, and the average number of bands was 4.5. In ISSR, the total number of bands was 30 with 10 primers, and the average number of bands was 3 (Table 4.11). Amplification bands showed genetic similarity between mother and in vitro raised plants. The presence of monomorphic bands in both

Table 4.9 Amplified RAPD primers from micropropagated plants from rachis

Primer	Primer sequences	Total no. of amplified bands	No. of scorable bands	Polymorphic bands (no.)	Range of amplification (bp)
OPE-03	CCAGATGCAC	80	8	0	350–1250
OPA-06	GGTCCCTGAC	30	3	0	350–950
OPE-12	TTATCGCCCC	40	4	0	350–1200
OPC-07	GTCCCGACGA	60	6	0	400–1200
OPP-10	TCCCGCCTAC	20	2	0	500–1000
OPH-04	GGAAGTCGCC	20	2	0	350–650
Total		250	25	0	

Source: Singh (2015) and Singh et al. (2016a, b, c)

RAPD and ISSR assays shows that use of in vitro leaves for regeneration purposes is safe and there is no adverse effect of in vitro conditions and culture media, as both donor plants and cultures have been maintained in in vitro conditions for longer duration. Combinations of both markers have been used by several researchers to assess the genetic stability in many micropropagated plants such as *Spilanthes acmella* (Yadav et al. 2014), *Musa acuminata* var. Nanjangud Rasabale (Lakshmanan et al. 2007), *Simmondsia chinensis* (Kumar et al. 2011), etc.

4.9.2 Genetic Fidelity in *Sapindus trifoliatus*

Genetic fidelity of in vitro raised plants from nodal explants (isolated from seedlings) was tested using 15 different primers of RAPD analysis (Table.4.12). All the tested 15 primers produced resolvable, reproducible and scorable bands. The average number of amplified bands was obtained 5.93/primer. Genomic DNA of mother plants (seedlings) and in vitro raised plantlets was used for the development of the RAPD profile. Amplified RAPD profile of mother plants and micropropagated plantlets showed similar banding pattern. All the tested primers produced monomorphic bands approving the genetic homogeneity of the tested plants. Polymorphism was not observed among the RAPD amplified profile of the micropropagated plants. RAPD-based molecular marker analysis revealed that there were no variations convinced during micropropagation. RAPD is a suitable molecular technique to detect the somaclonal variation that is induced or occurs during the micropropagation of plant species (Shu et al. 2003). A number of workers used RAPD to test the clonal fidelity of in vitro raised plants, viz., *Prunus dulcis* (Martins et al. 2004), banana (Venkatachalam et al. 2007), etc.

Table 4.10 Amplified ISSR primers from micropropagated plants from rachis

ISSR primer	Nucleotide sequence (5′–3′ motif)	Annealing Temp (°C)	Amplified bands (Total no.)	Scorable bands (no.)	Polymorphic bands (no.)	Amplification range (bp)
UBC-824	TCTCTCTCTCTCTCTCG	55	10	1	0	800
UBC-834	AGAGAGAGAGAGAGAGACYT	55	40	4	0	250–900
UBC-856	ACACACACACACACACCTA	56	20	2	0	400–1000
UBC-825	ACACACACACACACACT	50	40	4	0	500–1150
UBC-814	CTCTCTCTCTCTCTCTA	55	20	2	0	450–950
UBC-807	AGAGAGAGAGAGAGAGT	55	40	4	0	400–900
UBC-810	GAGAGAGAGAGAGAGAT	55	20	2	0	400–1100
UBC-840	GAGAGAGAGAGAGAGAY	58	30	3	0	400–1250
UBC-842	GAGAGAGAGAGAGAGAYG	59.1	30	3	0	450–950
UBC-854	TCTCTCTCTCTCTCTCAGG	58	30	3	0	700–1250
Total			280	28		

Source: Singh (2015) and Singh et al. (2016a, b, c)

Table 4.11 RAPD and ISSR primers amplified for in vitro raised plants from IVL explants

RAPD primer	Amplified bands (total no.)	Scorable bands (no.)	ISSR primer	Amplified bands (total no.)	Scorable bands (no.)
OPE-03	80	8	UBC-824	20	2
OPA-06	30	3	UBC-834	40	4
OPE-12	50	5	UBC-856	20	2
OPC-07	60	6	UBC-825	40	4
OPP-10	20	2	UBC-814	30	3
OPH-04	30	3	UBC-807	40	4
			UBC-810	20	2
			UBC-840	30	3
			UBC-842	30	3
			UBC-854	30	3
Total	270	27	Total	300	30

Table 4.12 Primers used for the assessment of genetic fidelity

Primer	Sequences (5′-3′)	No. of scorable bands	Primer	Sequences (5′-3′)	No. of scorable bands
OPAF-2	CAGCCGAGAA	6	OPAF-6	CCGCAGTCTG	8
OPAF-11	ACTGGGCCTC	7	OPAF-15	CACGAACCTC	7
OPAF-18	GTGTCCCTCT	3	OPAG-11	TTACGGTGGG	4
OPAG-19	A GCCTCGGTT	7	OPC-18	TGAGTGGGTG	6
OPL-20	TGGTGGACCA	3	OPN-16	AAGCGACCTG	5
OPO-6	CCACGGGAAG	7	OPY-15	AGTCGCCCTT	10
OPZ-11	CTCAGTCGCA	5	OPZ-17	CCTTCCCACT	3
OPZ-19	GTGCGAGCAA	8			

Source: Asthana et al. (2011)

4.10 Solubilizing Property

Pelegrini et al. (2008) and Nakayama et al. (1986) reported that *S. mukorossi* saponin and bisdesmoside from pericarps of *S. mukorossi* greatly enhanced the water solubilities of the co-occurring monodesmoside.

4.11 Biodiesel Production

Biodiesel can be defined as vegetable oil or animal fat-based diesel fuel which contains long-chain alkyl and esters. Production of biodiesel is a chemical reaction in which lipids react with alcohol producing fatty acid esters (Wikipedia 2018).

Soapnut oil has been used for production of biodiesel, and soapnut oil methyl esters (SNME) had satisfactory fuel properties (Chen et al. 2013). They reported that the predominant components were methyl oleate at 55.2% and methyl eicosenoate at 23.9% and have high amount of monounsaturated fatty acid methyl esters at 79.1% by weight. The biodiesel produced from the soapnut oil has shown all the characteristic features of biofuel such as cetane number (58), cold filter plugging point of 6 °C, flash point of 177 °C, iodine value of 83.6 g I_2/100 g and kinematic viscosity of 4.88 mm^2/s at 40 °C. Additionally, the oxidation stability of the SNME fulfilled the requirements of the biodiesel specifications. This biodiesel was blended with diesel at various volumetric percentages and also evaluated for fuel properties. Some other nonedible oil (soapnut and Jatropha oil) was also used for the production of biodiesel (Chhetri et al. 2008).

Questions

1. Who coined the term biotechnology?
2. What is the means of organogenesis and somatic embryogenesis and point out the differences between these two terms?
3. Define the term genetic fidelity and its use in plant biotechnology.
4. What is micropropagation and explain its advantages?
5. How many nucleotides may present in primers of RAPD?
6. How many nucleotides may present in primers of ISSR?
7. What is biodiesel and what are its advantages over conventional diesel?

Further Readings

Asthana. (2010). PhD thesis, Banaras Hindu University, Varanasi, UP, India.
Singh R. (2015). PhD thesis, Banaras Hindu University, Varanasi, UP, India.
Singh, R., Rai, M. K., & Kumari, N., (2015). Somatic embryogenesis and plant regeneration in *Sapindus mukorossi* Gaertn. from leaf-derived callus induced

with 6-Benzylaminopurine. *Applied Biochemistry and Biotechnology, 177*, 498–510.

Singh, R., Kashyap, S. P., Kumari, N., & Singh, M. (2016). Regeneration of soapnut tree through somatic embryogenesis and assessment of genetic fidelity through ISSR and RAPD markers. *Physiology and Molecular Biology of Plants, 22*, 381–389.

Phytochemical Analysis and Pharmaceutical Development from *Sapindus* spp.

5

Plants serve as the source for many useful things for living being's food, fibres, medicines, etc. Their utility in treatment of various diseases has been identified since ancient time. Green plants produce and preserve a range of biochemical products. Large number of phytochemicals is extractable and found use as chemical feedstocks/raw material for numerous scientific research. A number of plant secondary metabolites are commercially important and used for the production of pharmaceutical compounds. Plant-based herbal products are offered as cost-effective raw materials for the development of modified derivatives with higher efficacy and lesser toxicity. Countless pharmaceutically important phytocompounds including vinblastine, vincristine, Taxol, podophyllotoxin, camptothecin, digitoxigenin, digoxigenin, tubocurarine, morphine, codeine, aspirin, atropine, pilocarpine, capsaicin, allicin, curcumin, artemisinin and ephedrine are available in the market either in pure form or in modified form. The chemical structures of these phytochemicals are shown in Fig. 5.1.

Sometimes, the medicinal plant's crude extract may be used as curatives. On the other hand, isolation, purification and identification of the active principles from the crude extracts and explanation of molecular mechanisms/pathways involved or mode of action of a phytocompound/drug are of utmost importance. Hence, to carry out extensive research to assess their best efficiency both in the mixture as a traditional medicine showing synergistic response or/and single active compound of better potential. To avoid the expensive chemical synthesis of a compound, the product must be obtained from the plant materials. Different steps such as isolation, characterization and identification of secondary metabolites produced by plants are required followed by their evaluation as therapeutics (Taylor et al. 2001).

R. Singh, B. Sharma, *Biotechnological Advances, Phytochemical Analysis and Ethnomedical Implications of Sapindus species*,
https://doi.org/10.1007/978-981-32-9189-8_5

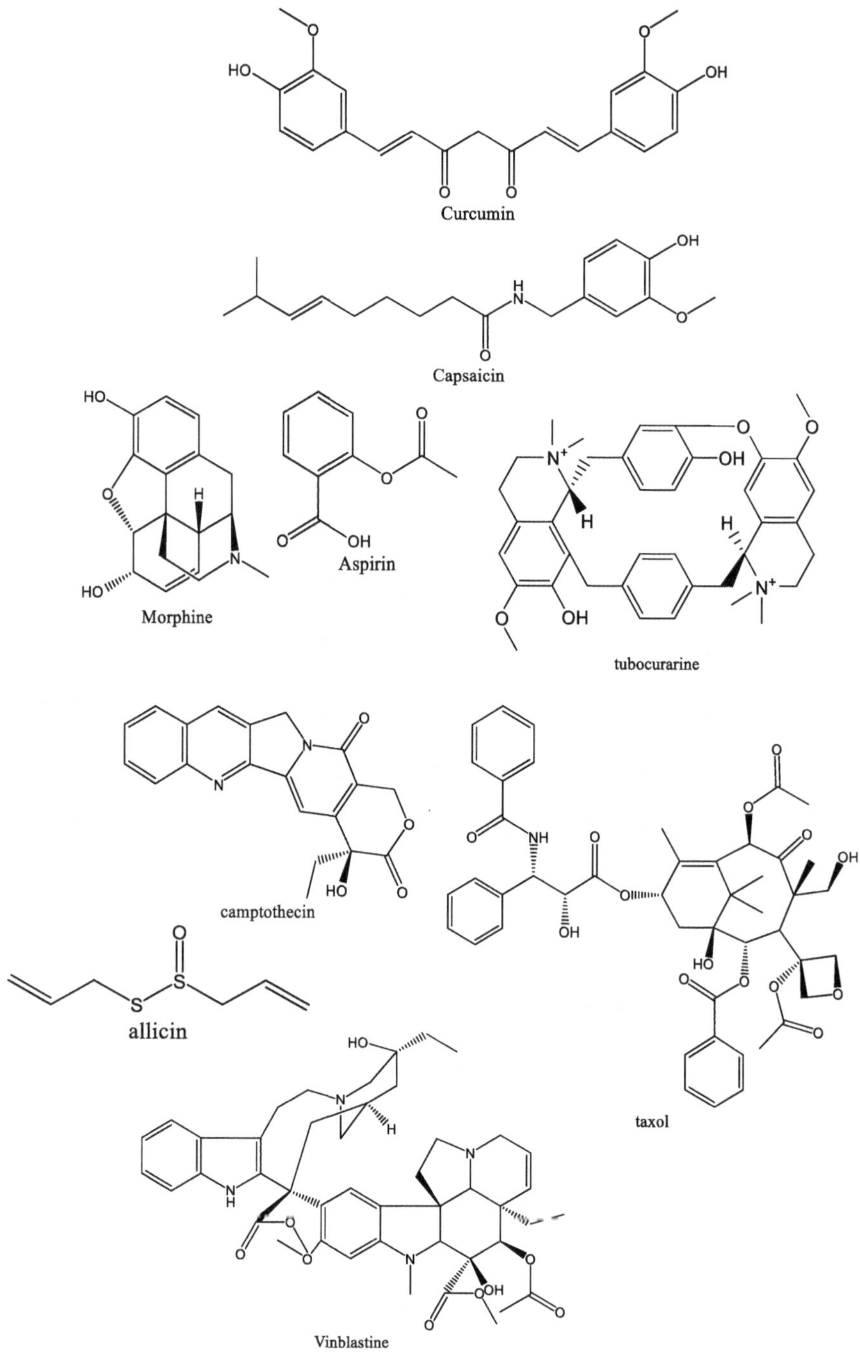

Fig. 5.1 Chemical structures of various plant-derived compounds

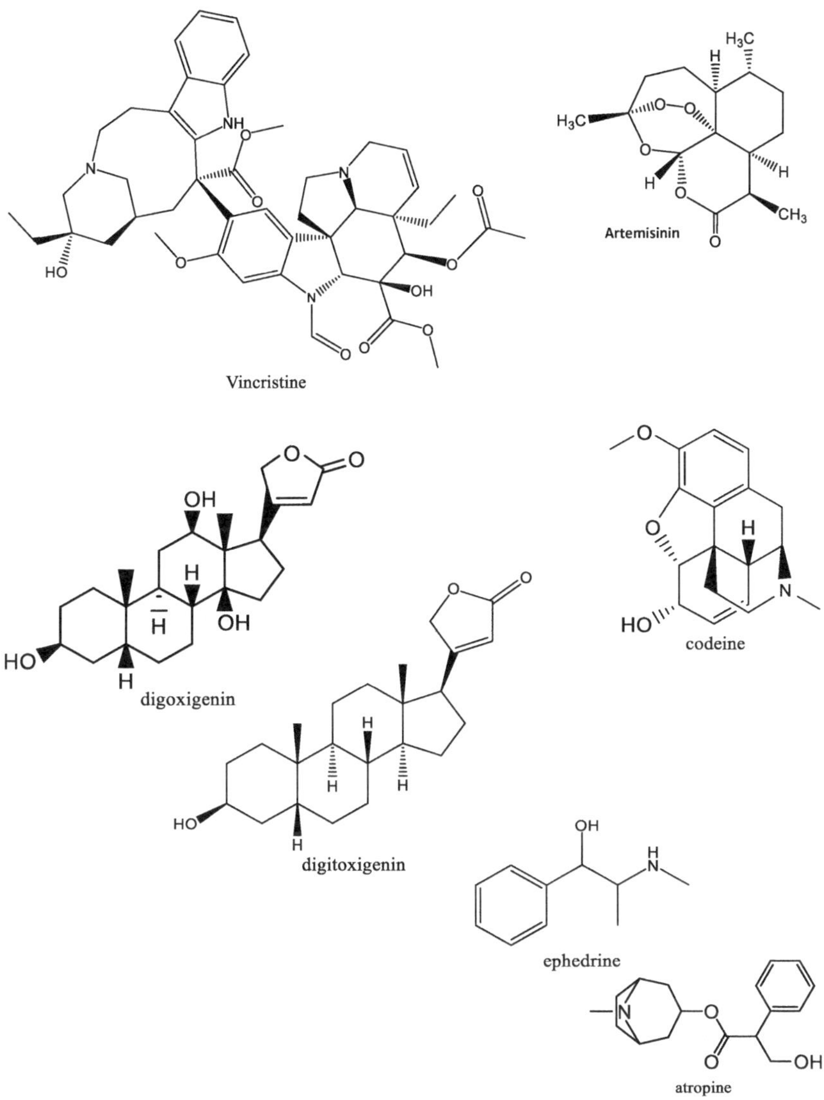

Fig. 5.1 (continued)

5.1 Phytochemistry of *Sapindus*

Genus *Sapindus* is well known for the high content of different types of saponins. Different types of saponins are present in fruits, roots and galls (Suhagia et al. 2011). The fruits of *S. mukorossi* contain six types of saponins, and it is valued for the saponins (10.1%) in the pericarp (Suhagia et al. 2011; Upadhyay and Singh 2012). Besides the saponins, *Sapindus* also contains variety of phytocompounds. Seeds of *S. mukorossi* contain 23% oil of which 92% is triglycerides (Suhagia et al. 2011). The triglyceride fraction contained 30% oleopalmito-arachidin glyceride, 13.3% oleo-diarachidin glyceride and 56.7% di-olein type glycerides such as dioleopalmitin, dioleo-stearin and dioleo-arachidin. The seed kernel cake contains 7.9% total N and 32% crude protein. Mainly globulin-type protein was present. The amino acids, namely, glycine, alanine, proline, tryptophan, lysine, serine, arginine, valine, leucine/isoleucine, aspartic acid and glutamic acid, have been identified. However, the kernels lack most of the essential amino acids (44%). This can be considered as poor supplement for human nutrition due to deficiency of essential amino acids (Orwa et al. 2009). But by adding essential amino acids, a complete industrial protein could be prepared from the globulin fraction. Almost every part of *S. mukorossi* contains a variety of phytocompouds; some of these are presented in Table 5.1.

5.2 Phytochemistry of *S. trifoliatus, S. saponaria* and *S. rarak*

Saponins content in *S .rarak* extract has been reported to be 7.5% (Hamburger et al. 2007), whereas in *S. saponaria* saponins contents are higher (12%) (Abreu et al. 2004). The chemical structure of saponins is almost similar in *S. rarak and S. saponaria* fruit's pericarps. *S. rarak and S. saponaria* have monodesmoside triterpenoid saponins that retain hederagenin as the aglycone (Hamburger et al. 2007; Lemos et al. 1992, 1994). Differences in saponins occur mainly due to change in sugar composition and arrangements of side chains. In *S. rarak,* arabinose is attached to hederagenin and rhamnose and xylose as sugar residues attached to arabinose (Hamburger et al. 2007). While in *S. saponaria*, glucose is directly attached to hederagenin, and rhamnose and arabinose are attached to the glucose (Lemos et al. 1992, 1994). A number of phytocompounds have been identified and isolated from whole plants, fruits, fruits peel, pericarp, etc. of *Sapindus* sps. (Table 5.2).

5.3 Saponins

Saponins are a group of naturally occurring glycosides that are primarily present in the plant kingdom and can be present in different parts of plant such as roots, tubers, shoots, leaves, flowers and seeds. Formation of soapy lather is characteristic feature of most of the saponins, when shaken in water solution (Oleszek 2000). On the basis of carbon skeleton, saponins are classified I to main category named triterpenes and steroids. About 100 of plant families contain different kinds of saponins.

Phytochemical studies of ethanolic and aqueous extracts of fruits of *S. mukorossi* showed the presence of alkaloids, phytosterols, phenolic compounds, tannins, flavonoids, glycosides and saponins (George and Shanmugam 2014). Of these all

Table 5.1 Various isolated phytocompounds from different parts of *S. mukorossi*

Plant's part	Chemical constituents/compounds isolated	References
Leaves	4-Allyl-2-methoxyphenyl-3-*O*-α-L-rhamnopyranosyl-(1→6)-β-D-glucopyranoside	Ni et al. (2006)
Roots	3-*O*-α-L-arabinopyranosyl-(1→3)-α-L-rhamnopyranosyl-(1→2)-[α-Larabinopyranosyl-(1→3)]-β-D-glucopyranosyl-21,23R-epoxyl tirucalla-7,24-diene-21β-ethoxy-3β–ol; 3-*O*-β-D-xylopyranosyl-(1→3)-α-L-rhamnopyranosyl-(1→2)-[β-L-arabinopyranosyl-(1→3)]-β-D-glucopyranosyl-21,23R-epoxyl tirucalla-7,24-diene-21β-ethoxy-3β–ol; 3-*O*-β-D-xylopyranosyl-(1→3)-α-L-rhamnopyranosyl-(1→2)-[α-L-arabinopyranosyl-(1→3)]-β-D-glucopyranosyl-21,23R-epoxyl tirucall-α-7,24-diene-21β-methoxy-3β–ol; 3-*O*-α-L-arabinopyranosyl-(1→3)-α-L-rhamnopyranosyl-(1→2)-[α-Lrhamnopyranosyl-(1→3)]-β-D-glucopyranosyl-21,23R-epoxyl tirucalla-7,24-diene-21β-ethoxy-3β–ol; 3-*O*-α-Larabinopyranosyl-(1→3)-α-L-rhamnopyranosyl-(1→2)-[α–L-rhamnopyranosyl-(1→3)]-β-D-glucopyranosyl-21,23R-epoxyl tirucalla-7,24-diene-21β-methoxy-3β-ol; 3-*O*-α-L-rhamnopyranosyl-(1→6)-β-D-glucopyranosyl-21,23R-epoxyl tirucalla-7,24-diene-21β-ethoxyl-3β-ol	Huang et al. (2005)
	3-*O*-α-L-rhamnopyranosyl-(1→2)-[α-L-arabinopyranosyl-(1→3)]-β-D-glucopyranosyl (21,23R)-epoxyl tirucalla-7,24-diene-(21S)-ethoxyl-3β-ol; 3-*O*-α-L-rhamnopyranosyl-(1→2)-[α-L-arabinopyranosyl-(1→3)]-β-D-glucopyranosyl (21,23R)-epoxyl tirucall-7, 24-diene-(21S)-methoxyl-3β-ol	Huang et al. (2007)
Fruit	Hederagenin3-*O*-α-L-arabinopyranosyl(2→1)-α-L-rhamnopyranosyl(3→1) xylopyranosyl(4→1)glucopyranosyl [(6→1)rhamnopyranosyl](2→1)glucopyranoside;	Chirva et al. (1973a)
	3-O-{α-L-arabinopyranosyl-(1→2)-α-L-arabinopyranosyl-(3→1)-α-L-rhamnopyranosyl}-hederagenin; 3-O-{β-D-xylopyranosyl-(1→2)-α-L-arabinopyranosyl-(3→1)-α-L-rhamnopyranosyl}hederagenin	Azhar et al. (1994)
	Eupha 7,24-dien-3-tetradecanoate; eupha 7,24-dien-3-pentadecanoate; eupha 7,24-dien-3-hexadecanoate; eupha,7,24-dien-3-heptadecanoate; eupha,7,24-dien-3-nonadecanoate;eupha,7,24-dien-3-heneicosanoate;	Huang et al. (2003)
	hederagenin3-O-α-L-arabinopyranosyl(2→1)-α-L-rhamnopyranosyl(3→1)-β-D- xylopyranosyl(4→1) glucopyranoside	Chirva et al. (1973c)
Pericarp	Hederagenin3-O-α-L-arabinopyranosyl(1→3)-α-L-rhamnopyranoside; Hederagenin3-O-α-L-arabinopyranosyl (1→3)-α-L-rhamnopyranosyl(1→2)-β-D-xylopyranoside	Chirva et al. (1970)
	hederagenin3-O-α-L-arabinopyranosyl(2→1)-α-L-rhamnopyranosyl(3→1)-β-D-xylopyranosyl(4→1) glucopyranosyl[(2→1)gluco-pyranoside] (5→1) rhamnopyranoside	Chirva et al. (1973b)
	hederagenin3-O-α-L-arabinopyranosyl(1→3)-α-L-rhamnopyranosyl(1→2)-α-L-arabinopyranoside; hederagenin3-O-α-L-arabinopyranosyl(1→3)-α-L-	Kimata et al. (1983)

(continued)

Table 5.1 (continued)

Plant's part	Chemical constituents/compounds isolated	References
	rhamnopyranosyl(1→2)-α-L-arabinofuranosyl;hederagenin3-O-α-L-arabinopyranosyl(1→3)-α-L-rhamnopyranosyl (1→2)-β-D-xylopyranosyl-4-acetyl ; Hederagenin3-O-(3,4-di-O-acetyl-β-D-xylopyranosyl)-(1→3)-α-L-rhamnopyranosyl-(1→2)-α-L-arabinopyranoside; Hederagenin,28-di-O-glycosides	
	1-O-α-L-rhamnopyranosyl-(1→2)-[α-L-rhamnopyranosyl-(1→3)]-β-D-glucopyranoside-12-O-α-L-rhamnopyranosyl-(1→2)β-D-glucopyranoside; hederagenin12-O-α-rhamnopyranosyl-(1→2)-β-glucopyranoside-1-O-α-rhamnopyranosyl-(1→2)-[α-rhamnopyranosyl-(1→3)]-β-glucopyranoside;hederagenin1,12-bisglycoside-O-α-L-rhamnopyranosyl-(1→2)-[α-L-rhamnopyranosyl-(1→3)]-β-D-glucopyranoside;hederagenin1,12-bisglycoside-O-α-rhamnopyranosyl-(1→2)-[α-rhamnopyranosyl-(1→3)]-β-lucopyranoside;1-O-α-L-rhamnopyranosyl-(1→2)-[α-rhamnopyranosyl-(1→3)]-β-D-glucopyranoside-12-hydroxy-all-trans-farnesol;1-O-α-rhamnopyranosyl-(1→2)-[α-rhamnopyranosyl-(1→3)]-β-glucopyranoside-12-hydroxyfarnesol; 12-hydroxy-all-trans-farnesol; 1-hydroxylinalool; GoshonosideF1; 8-hydroxygeraniol; β-glucosidegeraniol; 11(S)(E,E)-2,6-dodecadiene-3,7,11-trimethyl-1,12-diol	Azhar et al. (1993)
	11(S)-2,6(E,E)-dodecadiene-3,7,11-trimethyl-1,12-diol-1,12-bis-O-α-L-rhamnopyranosyl-(1→2)-[α-L- rhamnopyranosyl-(1→3)]-6-O-acetyl-β-D-glucopyranoside	Kuo et al. (2005)
	Hederagenin3-O-(2,4-O-Di-acetyl-α-L-arabinopyranoside)-(1→3)-α-L-rhamnopyranosyl-(1→2)-α-L-arabinopyranoside; Hederagenin3-O-(3,4-O-Di-acetyl-α-L-arabinopyranoside-(1→3)-α-L-rhamnopyranosyl-(1→2)-α-L-arabinopyranoside; Hederagenin3-O-α-L-Arabinopyranoside	Sengupta et al. (1975)
Galls	3β,7β,20(S),22-tetrahydroxydammar-24-ene-3-O-α-L-rhamnopyranosyl-(1→2)-β-D-glucopyranoside; 3β,7β,20 (S),22,23-pentahydroxydammar-24-ene-3-O-α–L-rhamnopyranosyl-(1→2)-β-D-glucopyranoside; 3β,7β,20 (S),22,25-pentahydroxydammar-23-ene-3-O-α-L-rhamnopyranosyl-(1→2)-β-D-glucopyranoside; 25-methoxy-3β,7β,20(S),22-tetrahydroxydammar-23-ene-3-O-α-L-rhamnopyranosyl-(1→2)-β-D-glucopyranoside; 25-methoxy-3β,7β,20(R)-trihydroxydammar-23-ene-3-O-α-L-rhamnopyranosyl-(1→2)-β-D-glucopyranoside; 4-Allyl-2-methoxyphenyl-6-O-α-L-arabinopyranosyl-(1→6)-β-D-glucopyranoside	Zikova and Krivenchuk (1994)
	21β-methoxy-3β,21(S),23-(R)-epoxytirucalla-7,24-diene-3-O-α -L-rhamnopyranosyl-(1→6)-β-D-glucopyranoside; 21α-methoxy-3β,21(R),23-(R)-epoxytirucalla-7,24-diene-3-O-α -L-rhamnopyranosyl-(1→6)-β-D-glucopyranoside; 21α -methoxy-3β,21-(R),23(R)-epoxytirucalla-7,24-diene-3-	Ni et al. (2004)

(continued)

Table 5.1 (continued)

Plant's part	Chemical constituents/compounds isolated	References
	O-α-L-rhamnopyranosyl-(1→2)-β-D-glucopyranoside; 21β-methoxy-3β,21(S),23(R)-epoxytirucalla-7,24-diene-3-O-α -L-dirhamnopyranosyl-(1→2,6)-β-D-glucopyranoside; 21α-methoxy-3β,21(R),23(R)-epoxytirucalla-7,24-diene-3-O-α-L-dirhamnopyranosyl-(1→2,6)-β-D-glucopyranoside	
Fruits and galls	Stigmasterol	Linde (1979)
	Hederagenin-3-O-(3-O-acetyl-α-L-arabinopyranosyl)-(1→3)-α-L-rhamnopyranosyl-(1→2)-α–Larabinopyranoside; hederagenin-3-O-(4-O-acetyl-α-L-arabinopyranosyl)-(1→3)-α-L-rhamnopyranosyl-(1→2)-α-L-arabino-pyranoside; hederagenin-3-O-(2,3-O-diacetyl-β-D-xylopyranosyl)-(1→3)-α-L-rhamnopyranosyl-(1→2)-α-L-arabinopyranoside; hederagenin-3-O-(2,4-O-diacetyl-β-D-xylopyranosyl)-(1→3)-α-L-rhamnopyranosyl-(1→2)-α-L-arabinopyranoside;hederagenin3,7,20(S)-trihydroxydammar-24-ene-3-O-α-L-rhamnopyranosyl-(1→2)-β-D-glucopyranoside; 3,7,20(R)-trihydroxydammar-24-ene-3-O-α-L-rhamnopyranosyl-(1→2)-β-D-glucopyranoside	Huang et al. (2006)
Fruit husk	Hederagenin; platycodigenin; saikogenin; oleanolic acid	Takagi et al. (1980)

reported compounds in *Sapindus* species, saponins were considered as the most active group that is most probably accountable for the various pharmacological activities (Francis et al. 2002; Goyal et al. 2014). These saponins are present in all parts of plant but are especially intense in the fruit (Upadhayay and Singh 2012; Sharma et al. 2011; Arora et al. 2012). *Sapindus* saponins are a mixture of six sapindosides (sapindosides A, B, C, D and mukorozisaponins (E1 and Y1), with sapindoside B considered a prime constituent and identified by advanced chromatographic and mass spectroscopic techniques, i.e. LC-MS (Saxena et al. 2004). Saponins are a large family of structurally related compounds of steroid or triterpenoid aglycone (sapogenin) attached to one or more oligosaccharide moieties by glycosidic bonds. One or more unsaturated C-C bonds may present in the aglycone or sapogenin. Normally, oligosaccharide chain is linked at the C3 position (monodesmosidic), but many saponins contain an additional sugar moiety at the C2,6 or C2,8 position (bidesmosidic). The carbohydrate moiety comprises the uronic acid, pentoses or hexoses. The variability and complexity of the saponin structure rise from the changeability of the aglycone structure, from nature of the side chains and by varying the position of these moieties on the aglycone (Francis et al. 2002). Saponins are difficult to classify due to the complexity and variability in structure. Vincken et al. (2007) classified the saponins on the basis of their biosynthesis. Various types of triterpene, saponins of oleanane, dammarane and tirucullane type have been isolated from the galls, fruits and roots of different *Sapindus* species. The structures of these saponins are demonstrated in Fig. 5.2. A brief account of saponins including their type and chemical names is shown in Table 5.3.

Table 5.2 Various phytocompounds isolated from *Sapindus* sps.

Sapindus species	Parts	Chemical constituents	References
S. trifoliatus	Pericarp	1,12-bis-*O*-β-D-xylopyranosyl(1→3)-[α-L-rhamnopyranosyl(1→2)]-β-D-glucopyranoside; 15-hydroxyfarnesol;15-*O*-β-xylopyranosyl-(1→3)-[α-rhamnopyranosyl (1→2)]-β-glucopyranoside farnesol; 15-*O*-β-xylopyranosyl(1→3)-[α-rhamnopyranosyl (1→2)]-β-glucopyronoside-15-hydroxyfarnesol; sesquiterpene oligoglycoside	Morikawa et al. (2009)
S. saponaria	Whole plant	3-*O*-{4-acetyl-β-D-xylopyranosyl-(1-3)-α-L-rhamnopyranosyl-(1-2)-α-L-arabnopyranosyl}-hederagenin;3-*O*-{α-L-arabinopyranosyl-(1-3)-α-L-rhamnopyranosyl-(1-2)-α-L-arabnopyranosyl}-hederagenin; 3-*O*-{β-D-xylopyranosyl-(1-3)-α-L-rhamnopyranosyl-(1-2)-α-L-arabnopyranosyl}-hederagenin	Gupta and Ahamad (1990)
	Fruit's peel	3-β-*O*-[α-L-rhamnopyranosyl-(1→3)-β-D-glycopyranosyl}-hederagenin	Ribeiro et al. (1995)
S. emarginatus	Fruits	Hederagenin-3-*O*-β-D-galactopyranosyl-(1→2)-β-D-glucopyranosyl; hederagenin-3-*O*-β-D-glucuronopyranosyl-(1→2)-β-D-galactopyranoside	Jain et al. (2009)
	Pericarp	Hederagenin-3-*O*-α-D-arabinopyranoside; hederagenin-3-*O*-(2-O-acetyl-β-D-xylopiranosyl)-(1→3)-α-L-rhamnopyranosyl-(1→2)-α-L-arabinopyranoside; 23-O-acetyl-hederagenin-3-*O*-(4-O-acetyl-β-D-xylopyranosyl)-(1→2)-α-L-rhamnopyranosyl-(1→2)α-L-arabinopyranoside; oleanolic acid 3-O-(4-O-acetyl-β-D-xylopyranosyl)-(1→2)-α-L-rhamnopyranosyl-(1→2)α-L-arabinopyranoside	Mahabusarakam et al. (1990)
		3-O-{β-D-xylopyranosyl)-(1→3)-α-L-rhamnopyranosyl-(1→2)α-L-arabinopyranosyl}hederagenin;3-O-{α-L-arabinopyranosyl(1→3)-α-L-rhamnopyranosyl-(1→2)α-L-arabinopyranosyl}; β-sitosterol	Kasai et al. (1988)
S. delavayi	Pericarp	Oleanolic acid 3-O-α-L-arabinopyranosyl -(1→3)-α-L-rhamnopyranosyl-(1→2)α-L-arabinopyranoside; monodesmoside; oleanolic acid 3-O-β-L-xylopyranosyl -(1→3)-α-L-rhamnopyranosyl-(1→2)α-L-	Lemos et al. (1992)

(continued)

Table 5.2 (continued)

Sapindus species	Parts	Chemical constituents	References
		arabinopyranoside; 3-O-β-D-xylopyranosyl-(1→3)-α-L-rhamnopyranosyl-(1→2)α-L-arabinopyranoside-3-acetyl; 3-O-α-L-arabinopyranosyl (1→3)-α-L-rhamnopyranosyl-(1→2)β-D-arabinopyranosyl-3,5-acetyl	
		1-O-α--rhamnopyranosyl-(1→2)-[α-L-rhamnopyranosyl-(1→3)]-β-glucopyranoside-12-hydroxyfarnesol; 1-O-α--rhamnopyranosyl-(1→2)-[α-L-arabinopyranosyl-(1→3)]-β-glucopyranoside-12-hydroxyfarnesol; 1,12-O-bisglycoside-α-rhamnopyranosyl-(1→3)-[α-L-rhamnopyranosyl-(1→2)]-α-arabinopyranosyl-(1→3)-[α-rhamnopyranosyl-(1→2)]; 1,12-O-bisglycoside-α-L-rhamnopyranosyl-(1→3)-[α-L-rhamnopyranosyl-(1→2)]-α-L-arabinopyranosyl-(1→3)-[α-rhamnopyranosyl-(1→2)]; 1,12-O-bisglycoside-α-L-rhamnopyranosyl-(1→3)-[α-L-rhamnopyranosyl-(1→2)]-α-L-arabinopyranosyl-(1→3)-[α-L-rhamnopyranosyl-(1→2)];	Lemos et al. (1994)

5.3.1 Biosynthesis of Saponins

Saponins are plant-based natural compounds which generally act against pathogens and herbivores.

In spite of their role in plant defence, saponins have been used in several traditional medicine systems and possess valuable pharmacological properties. Hundreds of plant species have been reported for the saponin production; a complete biosynthetic pathway for saponins has not been known in any plant species. In current scenario, it is supposed that intermediates of the phytosterol pathway are a major precursor for saponins. It is supposed that there are several enzymes involved in this pathway mainly belonging to the multigene families of oxidosqualene cyclases (OSCs), cytochromes P450 (P450s) and family 1 UDP-glycosyltransferases (UGTs). About 150 types of different saponins have been discovered. The unique structural features involve advance biosynthetical enzymes of different phylogenetic families such as in oats synthesis of triterpenoid saponins involved a serine carboxypeptidase-like acyltransferase (SCPL) (Augustin et al. 2011).

Acetyl-CoA enters through the mevalonate pathway and is finally converted into the 2,3-oxidosqualene into the multiple steps. The phytosterols synthesis pathway, phytosterols and triterpenoid saponins shared the common precursors

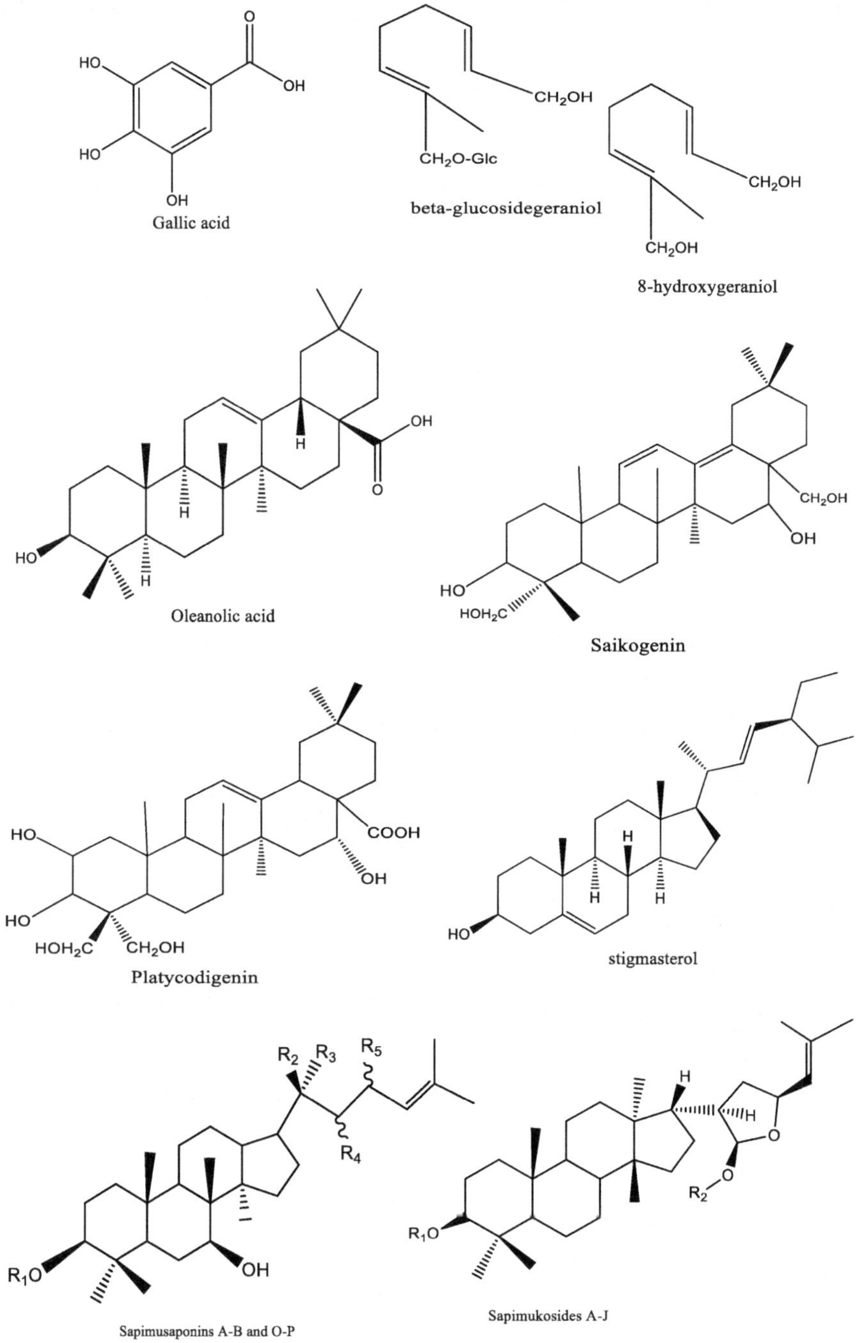

Fig. 5.2 Chemical structures of phytocompounds of *Sapindus* species

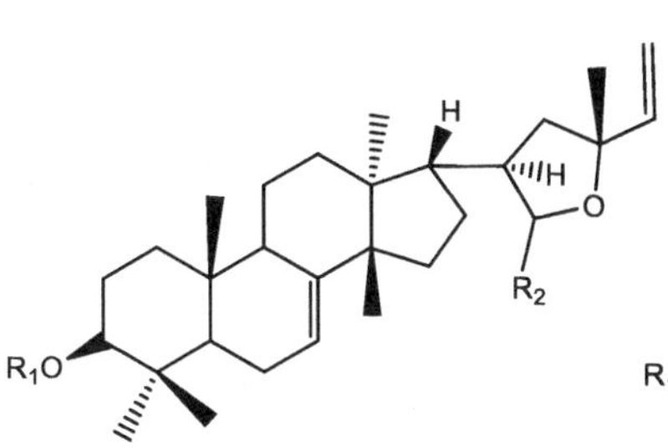

Sapimusaponins F-J and Q-R

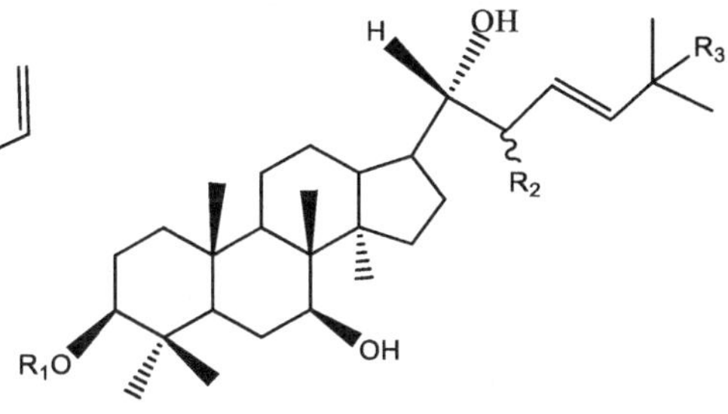

Sapimusaponins C-E

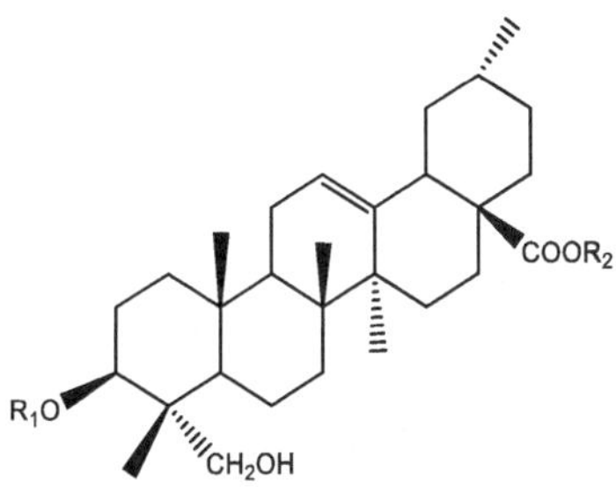

Sapimusaponins K-N, Sapindosides A-E, mukorozi saponin E1

Fig. 5.2 (continued)

2, 3-oxidosqualene. The portion of the triterpenoid saponin synthesis pathways along with synthesis of phytosterols is described in Fig. 5.3a. This concept is supported by the reported increased level of sterols and saponins in ginseng plants (*P. ginseng*) overexpressing squalene synthase (Lee et al. 2004). Terpenoid biosynthesis in plants is extensively classified, and triterpenes such as steroids and saponins are primarily produced in the cytosol utilizing IPP through the mevalonate pathway (Chappell 2002; Kirby and Keasling 2009; Trojanowska et al. 2000). All terpenoids derive from condensation of five-carbon building blocks designated IPP (3-isopentenyl pyrophosphate, C5) and DMAPP (dimethylallyl pyrophosphate, C5). In plants IPP and DMAPP are derived from the two pathways, i.e. through the cytosolic mevalonate pathway from the condensation of acetyl-CoA and through the plastidial MEP pathway from pyruvate and phosphoglyceraldehyde. GPP (geranyl pyrophosphate, C10) is produced from the condensation of IPP and DMAPP, and addition of a second IPP unit into the GPP leads to FPP (farnesyl pyrophosphate, C15), the mutual precursor of the wide range of sesquiterpenes produced by plants. Squalene (C30) is formed by the linkage of two FPP units which subsequently is converted to 2,3-oxidosqualene (C30). Phytostrols, triterpenoid saponins and steroidal saponins are synthesized from the common

Table 5.3 Different types of saponins isolated from the genus *Sapindus*

Saponins	Type	Chemical name	References
Sapindoside			
A	Oleanane	Hederagenin-3-*O*-α-L-arabinosyl- (2→1)-α-L-rhamnopyranoside	Chirva et al. (1970)
B		Hederagenin-3-*O*-α-L-arabinosyl- (2→1)-*O*-α-L-rhamnopyranosyl-(3→1)-β-D-xylopyranoside	
C		Hederagenin-3-*O*-β-D-glucosyl(1→4)- β-D-xylosyl (1→3)-α-Lrhamnosyl(1→2)- α-L-arabinoside	Chirva et al. (1970)
Sapinmukoside			
A	*Tirucullane*	3-*O*-α-L-rhamnopyranosyl-(1→2)-*α-L-arabinopyranosyl-(1→3)+–β-D-glucopyranosyl-21, 23R-epoxyl tirucall-7, 24R-diene-3 β, 2-diol	Teng et al. (2003)
B		3-*O*-α-L-rhamnopyranosyl-(1→6)-β-D-glucopyranosyl-21, 23R-epoxyl tirucall-7, 24R-diene-3 β, 21-diol	
C		3-*O*-α-L-rhamnopyranosyl-(1→2)- [α-L-arabinopyranosyl-(1→3)]-β-D-glucopyranosyl (21,23R)-epoxyl tirucalla-7, 24-diene-(21S)-ethoxyl-3β-ol	Ni et al. (2004)
D		3-*O*-α-L-rhamnopyranosyl-(1→2)- [α-L-arabinopyranosyl-(1→3)]-β-D-glucopyranosyl (21,23R)-epoxyl tirucall-7, 24-diene-(21S)-methoxyl-3β-ol	
E		3-*O*-α-L-arabinopyranosyl-(1→3)-α-L-rhamnopyranosyl-(1→2)-[α-L-arabinopyranosyl-(1→3)]-β-D-glucopyranosyl (21,23R)-epoxyl tirucalla-7,24-diene-21β-ethoxyl- 3β-ol	Ni et al. (2006)
F		{3-*O*-β-D-xylanopyranosyl-(1→3)-α-L-rhamnopyranosyl-(1→2)-[β-L-arabinopyranosyl-(1→3)]-β-D-glucopyranosyl 21,23R-epoxyl tirucalla-7, 24-diene-21β-ethoxyl-3β-ol}	
G		{3-*O*-β-D-xylanopyranosyl-(1→3)-α-L-rhamnopyranosyl-(1→2)-[α-L-arabinopyranosyl-(1→3)]-β-D-glucopyranosyl (21,23R)-epoxyl tirucalla-7,24-diene-21β-methoxy-3β-ol}	
H		{3-*O*-α-L-arabinopyranosyl-(1→3)-α-L-rhamnopyranosyl-(1→2)-[α-L-rhamnopyranosyl-(1→3)]-β-D-glucopyranosyl 21,23R-epoxyl tirucalla-7,24-diene-21β-ethoxy-3β-ol}	
I		{3-*O*-α-L-arabinopyranosyl-(1→3)-α-L-rhamnopyranosyl-(1→2)-[α-L-rhamnopyranosyl-(1→3)]-β-D-glucopyranosyl 21,23R-epoxyl tirucalla-7,24-diene-21β-methoxy-3β-ol}	
J		{3-*O*-α-L-rhamnopyranosyl-(1→6)-β-D-glucopyranosyl 21,23R-epoxyl tirucalla- 7,24-diene-21β-ethoxyl-3β-ol}	

(continued)

Table 5.3 (continued)

Saponins	Type	Chemical name	References
Sapinmusaponin			
A	Dammarane	3,7,20(S),22-tetrahydroxydammar-24-ene-3-*O*-a-L-rhamnopyranosyl-(1→2)-D- glucopyranoside	Yao et al. (2005)
B		3,7,20(S),22,23-pentahydroxydammar-24-ene-3- *O*-a-L-rhamnopyranosyl-(1→2)-D glucopyranoside	
C		3,7,20(S),22,25-pentahydroxydammar-23- ene-3-*O*-a-L-rhamnopyranosyl-(1→2)-D-glucopyranoside	
D		25-methoxy-3,7,20(S),22- tetrahydroxydammar-23-ene-3-*O*-a-L-rhamnopyranosyl-(1→2)-D-glucopyranoside	
E		25-methoxy-3,7,20(R)-trihydroxydammar- 23-ene-3-*O*-a-L-rhamnopyranosyl-(1→2)-D-glucopyranoside	
F	Tirucullane	21 β-methoxy-3-β-21(S), 23I-epoxy tirucall-7,24-diene-3-*O*-α-L-rhamnopyranosyl- (1→6)-β-D-glucopyranosyl	Huang et al. (2006)
G		21α-methoxy-3-β-21(S), 23I-epoxy tirucall-7,24-diene-3-*O*-α-L-rhamnopyranosyl-(1→6)-β-D-glucopyranosyl	
H		21α-methoxy-3-β-21(S), 23I-epoxy tirucall-7,24-diene-3-*O*-α-L-rhamnopyranosyl-(1→2)-β-D-glucopyranosyl	
I		21β-methoxy-3-β-21(S), 23I-epoxy tirucall-7,24-diene-3-*O*-α-L-dirhamnopyranosyl-(1→2,6)-β-D-glucopyranosyl	
J		21α-methoxy-3-β-21(S), 23I-epoxy tirucall-7,24-diene-3-*O*-α-L- dirhamnopyranosyl-(1→2,6)-β-D-glucopyranosyl	
K	Oleanane	Hederagenin-3-*O*-(3-*O*-acetyl-alpha-L-arabinopyranosyl)-(1→3)-alpha-L-rhamnopyranosyl-(1→2)-alpha-L-arabinopyranoside	Huang et al. (2005)
L		Hederagenin-3-*O*-(4-*O*-acetyl-alpha-L-arabinopyranosyl)-(1→3)-alpha-L-rhamnopyranosyl-(1→2)-alpha-L-arabino-pyranoside	
M		Hederagenin-3-*O*-(2,3-*O*-diacetyl-beta-D-xylopyranosyl)-(1→3)-alpha-L-rhamnopyranosyl-(1→2)-alpha-L-arabinopyranoside	
N		Hederagenin-3-*O*-(2,4-*O*-diacetyl-beta-D-xylopyranosyl)-(1→3)-alpha-L-rhamnopyranosyl-(1→2)-alpha-L-arabinopyranoside	
O	*Dammarane*	3,7,20(S)-trihydroxydammar-24-ene-3-*O*-alpha-L-rhamnopyranosyl-(1→2)-beta-D-glucopyranoside	
P		3,7,20(R)-trihydroxydammar-24-ene-3-*O*-alpha-L-rhamnopyranosyl-(1→2)-beta-D-glucopyranoside	
Q	Tirucullane	21α-methoxy-3β, 21I, 23(S)-epoxytirucall- 7,24-diene-3-*O*-β-D-glucopyranosyl-(1→2)- β-D-glucopyranoside	Huang et al. (2007)
R		21α-methoxy-3β, 21I, 23(S)-epoxytirucall- 7,24-diene-3-*O*-α-L-rhamnopyranosyl-(1→6)-β-D-glucopyranosyl-(1→2)-β-D- glucopyranoside	

(continued)

Table 5.3 (continued)

Saponins	Type	Chemical name	References
Mukorozisaponin			
G	Oleanane	Hederagenin-3-*O*-(2-*O*-acetyl-β-D-xylanopyranosyl)-(1→3)-α-L-rhamnopyranosyl-(1→2)-α-L- arabinoside	Huang et al. (2005)
E1		Hederagenin-3-*O*-α-L-arabinosyl- (1→3)-α-L-rhamnopyranosyl-(1→2)-α- L-arabinoside	Huang et al. (2005)

precursor, i.e. 2,3-oxidosqualene (e.g. Kalinowska et al. 2005; Phillips et al. 2006; Vincken et al. 2007).

5.3.2 Development of Sapogenin Diversity

Squalene epoxidase (SE) is a microsomal enzyme catalyzing the conversion of squalene into 2,3-oxidosqualene, which is successively cyclized by squalene epoxide cyclase (SEC) to form the phytosterols. During the cyclization process, internal bonds are introduced into the oxidosqualene backbone and formed primarily polycyclic molecules comprising variable numbers of five- and six-membered rings. Cyclization process offers the opportunity to establish the diverse internal linkages and produce a wide range of diverse structures, and more than 100 different triterpene skeletons (Xu et al. 2004) have been found in nature. However, only a few possible cyclization products of this diversity have been utilized in biosynthesis of saponin (Fig. 5.3b, Vincken et al. 2007).

5.4 Evaluation of Phytochemical Activity from *S. mukorossi*

5.4.1 Experimental Setup

5.4.1.1 Material Collection and Preparation of Extracts

Leaves and mature unripen fruits were collected from Banaras Hindu University, Varanasi, in April and June, respectively. All the plant material was collected from the same tree to maintain the genetic makeup and to avoid the other variability. Running tap water washing was used to remove the dust from plant materials. Fruit's pericarp and leaves were dried under shade condition and oven dried at 40–45 °C for 2 h to make coarse powder. Leaf and fruits powder was extracted using Soxhlet apparatus in ethanol, methanol and water. Extracts were then dried at 40 °C in rotary evaporator and stored −20 °C for further use. Percentage yield {PY, expressed in (w/w)} of crude extract was calculated using the following formula:

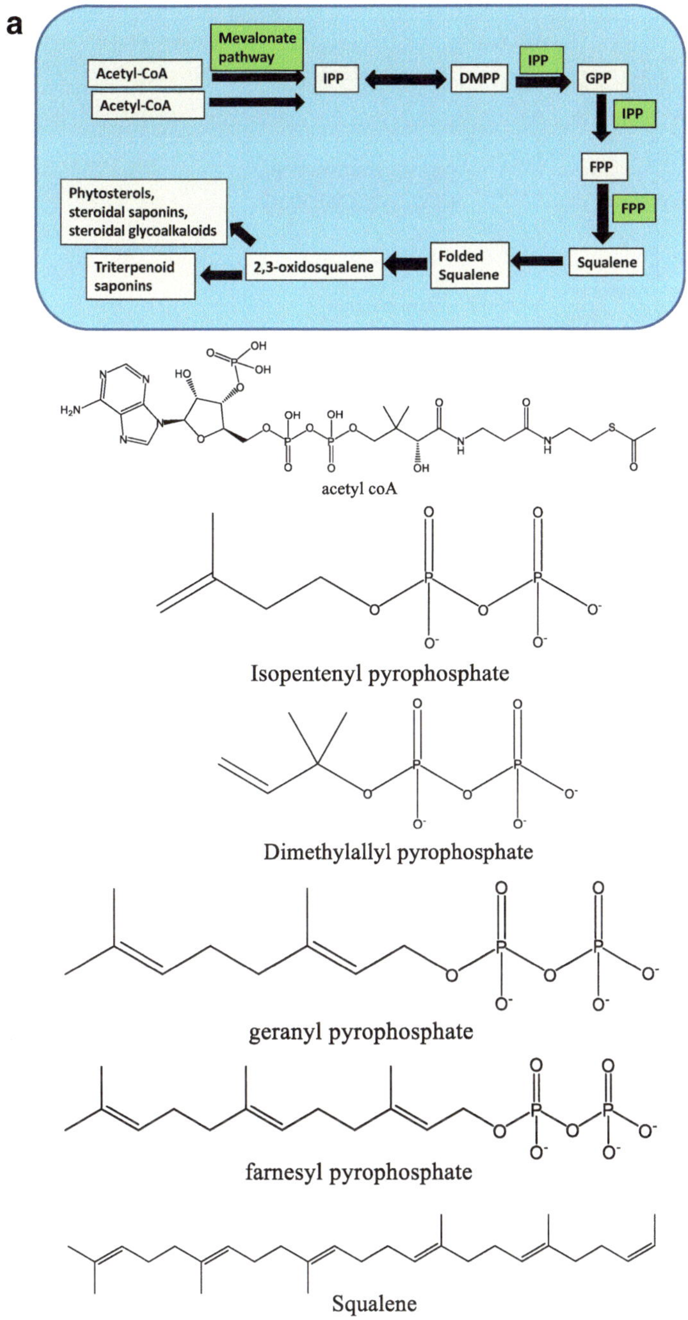

Fig. 5.3a Biosynthesis of triterpenoid saponins (*Acetyl-CoA* acetyl coenzyme A, *IPP* isopentenyl pyrophosphate, *DMAPP* dimethylallyl pyrophosphate, *GPP* geranyl pyrophosphate, *FPP* farnesyl pyrophosphate)

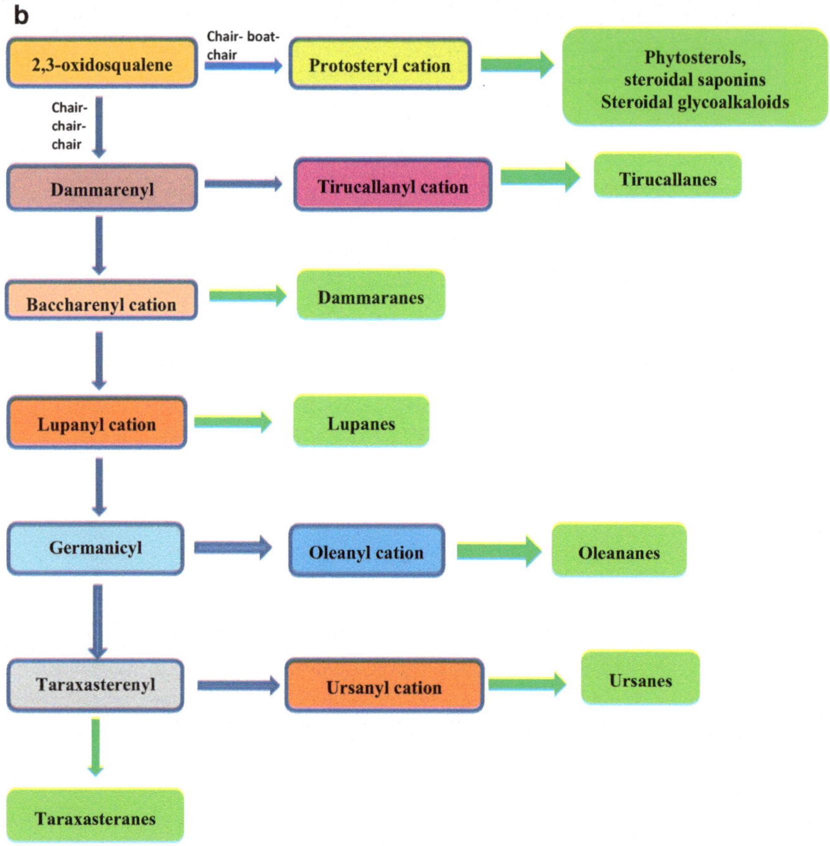

Fig. 5.3b Biosythesis pathway of different triterpenoid sapogenins

$$\mathrm{PY} = \frac{\text{Wt of crude extract recovered}}{\text{Wt of powder used}}$$

5.4.1.2 Preparation of Stock Sample of Extracts

Stock solution of 100 mg ml^{-1} concentration was obtained by dissolving each extract (1 g) in 10 ml of respective extraction solvents. According to the experiment's requirements, test samples were diluted in different concentrations.

5.4.1.3 Antioxidant Activity

(i) **Antioxidant activity using thin-layer chromatography**

Antioxidant activity of extracts was determined according to Tepe et al. (2005) with some alterations. Plant extract (4.54 mg was dissolved in the ratio of 1:10 in methanol) was applied on TLC plate and was developed in methanol and ethyl acetate (1:1).The plate was sprayed with 0.2% DPPH (prepared in methanol) and was left for 30 min at room temperature.

(ii) **DPPH-based antioxidant assay**

1,1-Diphenyl-2-picryl-hydrazyl (DPPH) is a stable free radical, in which one atom of nitrogen bridge contains an unpaired valence electron (Eklund et al. 2005). The free radical scavenging activity of the extracts was evaluated using DPPH assay (Yu et al. 2008). In brief, 1 ml sample (100–1200 μg ml^{-1}) of PE was added to DPPH solution (3 ml) (0.004%, dissolved in methanol) and shaken vigorously. The reaction mixture was incubated for 15 min. at room temperature. The absorbance of the samples was recorded at 517 nm using spectrophotometer (UV1, Thermo Scientific, US) and compared with ascorbic acid (standard). The result was expressed as the percentage inhibition of free radicals by the plant samples. Percentage of inhibition was calculated using the formula:

$$\text{DPPH free radical scavenging activity}(\%) = \frac{(A - B)}{(A)} \times 100$$

A = absorbance of control, B = absorbance of sample

(iii) **Antioxidant activity through β-carotene/linoleic acid assay**

Antioxidant activity was measured by using β-carotene and linoleic acid bleaching method (Miller 1971) with some modifications. Chloroform was used for preparation of stock solution of β-carotene (2 mg ml^{-1}). In a round-bottom flask, β-carotene solution (1 ml), linoleic acid (20 μl) and 200 μl of Tween 80 were mixed properly. Mixture was gently heated for complete removal of chloroform. To form an emulsion, 50 ml of water was added to the residue with vigorous agitation. Emulsion and extract were mixed in the ratio of 6:1 or 3:1 (2400 μl emulsion with 400 μl or 800 μl extract) in the test tube. Reaction mixture was incubated at 50 °C for 2 h, and absorbance was recorded immediately after incubation. DMSO served as control. Absorbance was taken at 470 nm. Percent inhibition was calculated as:

$$I\,(\%) = \frac{(\text{Absorbance } \beta - \text{caroteneafter 2 hassay})}{(\text{Absorbance of } \beta - \text{carotene initial})} \times 100$$

5.4.1.4 Estimation of Reducing Potential

Reducing potential of the extract was determined using the method of Nabavi et al. (2008) with some modifications. In brief, 2.5 ml of each, phosphate buffer solution (PBS, 0.2 M, pH-6.6), PE (100–500 μg ml^{-1}) and potassium ferricyanide (1%), was mixed. The above reaction mixture was incubated at 50 °C for 20 min. 2.5 ml trichloroacetic acid (TCA, 10%) was added to the reaction mixture to stop the reaction and centrifuged at 3000 rpm for 10 min. Supernatant (2.5 ml) was taken out and mixed with double distilled (2.5 ml) water and 0.5 ml ferric chloride solution. Ascorbic acid served as standard, and absorbance was recorded at 700 nm.

5.4.1.5 Thiobarbituric Acid Reactive Substance (TBARS) Assay

Lipid peroxidation inhibition was evaluated by modified TBARS method using egg yolk homogenates as lipid-rich media (Ohkawa et al. 1979). Malondialdehyde (MDA) is a secondary product of the oxidation of polyunsaturated fatty acids which reacts with two molecules of thiobarbituric acid (TBA). MDA yields a pinkish red chromogen having maximum absorbance at 532 nm (Janero 1990). Reaction mixture was prepared by following the method of Singh and Kumari (2015). In brief, 250 μl egg homogenate (10% v/v in DDW,) and 50 μl of extract were mixed properly in a test tube. 500 μL volume was maintained by adding DDW, and then 25 μl$FeSO_4$ (0.07 M) was added and incubated at room temperature for 30 min to persuade lipid peroxidation. After that, 750 μl acetic acid (20% v/v, pH 3.5) and 750 μl of 0.8% TBA (w/v, prepared in1.1%sodiumdodecyl sulphate) and 25 μl of 20% (w/v) TCA were added and vortexed. Final reaction mixture was incubated for 1 h in boiling water bath. After cooling, 1-butanol (3.0 ml) was added to each tube and centrifuged at 3000 rpm for 10 min. Supernatant (organic upper layer) was separated, and absorbance was measured at 532 nm.

5.4.1.6 Hydrogen Peroxide (H_2O_2) Scavenging Assay

The capacity of plant extracts to scavenge H_2O_2 was estimated by the method of Ruch et al. (1989) with some modifications. Plant extract was prepared in double distilled water. Hydrogen peroxide (80 μM) solution was prepared using phosphate buffer (50 mM, pH 7.4). Plant extract 3 ml (50 μg ml^{-1}) was added in H_2O_2 solution (0.6 ml, 80 μM). Reaction mixture was incubated for 50 min at room temperature, and absorbance was measured at 230 nm. Phosphate buffer used as blank and H_2O_2 solution served as control. H_2O_2 scavenging activity was calculated using the following formula:

$$\text{Hydrogenperoxide scavenging activity }(\%) = \frac{(Ai - At)}{(Ai)} \times 100$$

where Ai = absorbance of control and At = absorbance of test sample.

5.4.1.7 Estimation of Total Phenolic Content

Estimation of (TPC) was performed using Folin-Ciocalteu assay (McDonald et al. 2001). In brief, 5 ml Folin reagent (1:10 diluted with DDW) was added to 0.5 ml (200 μg ml^{-1}) PE, and finally 4 ml (1 M) aqueous sodium carbonate (Na_2CO_3) was added to this reaction mixture and incubated at room temperature for 15 min. Absorbance was recorded at 765 nm. Gallic acid was dissolved in methanol, and DDW (1:1) served as standard. TPC was expressed in terms of gallic acid equivalent (GAE, mg g^{-1} of dry mass), which is a common reference phenolic compound.

5.4.1.8 Estimation of Total Flavonoid Content

TFC was estimation by a reported method (Zhu et al. 2010) with slight modifications. PE (2 ml, 200 μg ml^{-1}) was added to 0.6 ml sodium nitrite (5%, w/v), 0.5 ml aluminium chloride (10%, w/v) and 3 ml sodium hydroxide (4.3%, w/v), and final volume (10 ml) was maintained with DDW. Finally reaction mixture was shaken for 6 min to complete the reaction. The reaction mixture was incubated at room temperature for 15 min. The absorbance was recorded at 500 nm. Rutin served as standard, and results were calculated as rutin equivalents (RE, mg g^{-1}).

5.5 Observations

Extract's yield was increased from methanol to ethanol and finally in double distilled water in both leaf and fruit. Extract's yield may be varied due to either polarity of solvents and/or various degrees of solubility of raw materials in different solvents. Percentage yield of fruit extract (48%, methanol; 59%, ethanol; 69%, aqueous) was higher than leaf extract (30%, methanol; 35%, ethanol; 39%, aqueous) in all used solvents. Higher yield was recorded in fruit when compared to leaf and may be due to the nature and composition of material, as after grinding, fruit powder was finer than leaf. This may also be due to high solubility of fruit powder in all solvents. The variation in yield may also be due to the polarity of the solvents, used in the extraction process.

5.5.1 Antioxidant Activity Through TLC and DPPH Assay

The discoloration of purple colour of DPPH solution on TLC plates proved the presence of antioxidant activity of the plant extracts.

The DPPH method is an easy, rapid, stable and sensitive approach to conclude the antioxidant activity (Koleva et al. 2002). DPPH is well known and strong radical which acts as a scavenger for other free radicals. All extracts have shown significant free radical scavenging action (Fig. 5.4). Antioxidant activity of fruit extract was higher in aqueous ($IC_{50} = 224.47$ μg ml^{-1}) than methanolic ($IC_{50} = 406.50$ μg ml^{-1}) and ethanolic ($IC_{50} = 1127.01$ μg ml^{-1}) extracts. In leaf, methanolic extract has shown higher scavenging activity ($IC_{50} = 307.05$ μg ml^{-1}) than aqueous ($IC_{50} = 395.80$ μg ml^{-1}) and ethanolic extracts ($IC_{50} = 1114$ μg ml^{-1}). Antioxidant potential is usually used as an important factor for testing the quality of medicinal bioactive and efficient components in food. Methanolic solution of DPPH is purple in colour and has characteristic absorption at 517 nm. When free radical scavenger removes the radicals (free odd electrons), the colour of DPPH solution turned light yellow from deep violet to light yellow, and absorption is reduced. The decrease in absorbance is the sign of increased in antioxidant activity of PE. Ascorbic acid served as a standard antioxidant and used as reference to determine the IC_{50} value of the PE (Sreekala et al. 2013; Nagvani et al. 2010).

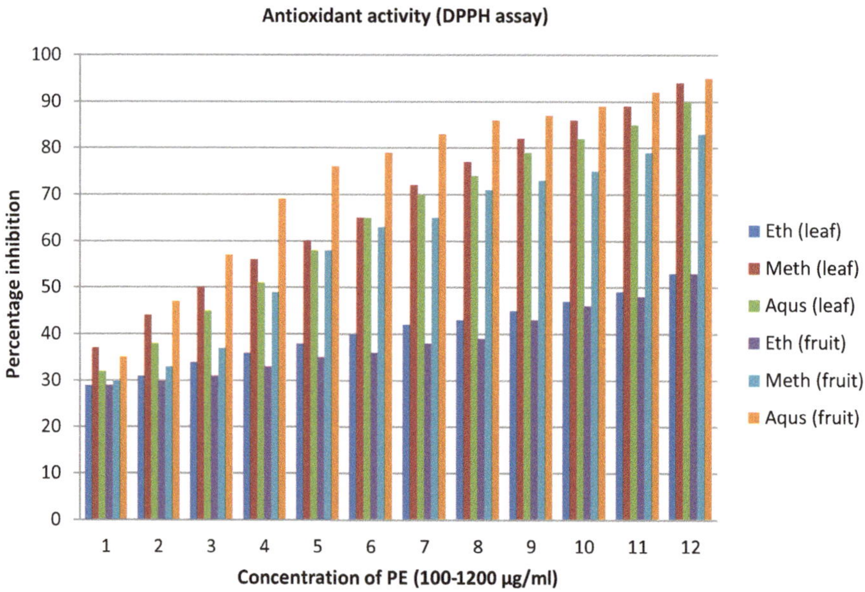

Fig. 5.4 Antioxidant activity of leaf and fruit extracts through DPPH method (IC_{50} of ascorbic acid is 5.00. IC_{50} was 1114 (ethanol, leaf), 307.05 (methanol, leaf), 395.80 (aqueous, leaf), 1127 (ethanol, fruit), 406.50 (methanol, fruit) and 224.47 (aqueous, fruit). All data is highly significant at $p \leq 0.001$)

5.5.2 β-Carotene/Linoleic Antioxidant Assay

This assay is considered as one of the most rapid approaches to test the antioxidant potential. In this assay, oxygenated water generates the reactive oxygen species (ROS); in result linoleic acid gets oxidized. All extracts of fruit and leaf have shown significant antioxidant activity, but leaf extract showed higher antioxidant activity than fruit extracts. All extracts, 800 µl volume showed higher antioxidant activity than 400 µl volume. In leaf, with 800 µl extract, highest antioxidant activity was recorded in methanolic extract followed by ethanolic and aqueous extract (Fig. 5.5).

In leaf extract (400 µl), highest antioxidant activity (80.07 ± 0.49%) was observed in aqueous extract, (60.1 ± 1.23%) in ethanolic and minimum (56.25 ± 1.38%) in methanolic extract. Fruit extract (800 µl) showed maximum free radical scavenging activity in aqueous extract followed by methanolic and ethanolic extract (Fig. 5.5), whereas with 400 µl (fruit extracts), maximum free radical scavenging activity (55.11 ± 0.69%) was recorded in aqueous extract, (39.30 ± 0.82%) in methanolic and minimum (30.64 ± 1.62%) in ethanolic extract. Thus, it is clear that antioxidant activity increased by increasing the concentration of plant extracts. Other workers has also used the β-carotene/linoleic assay for evaluation of antioxidant activity in several plant extracts (Lai and Lim 2011; Maisarah et al. 2013). In *Carica papaya,* maximum antioxidant activity was recorded in unripe fruit followed by young leave, ripe fruit and the seed through β-carotene bleaching assay (Maisarah et al. 2013).

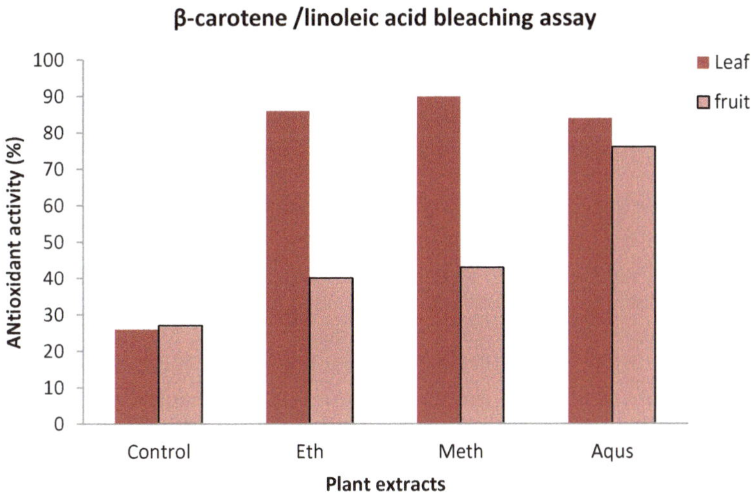

Fig. 5.5 Antioxidant activity through β-carotene/linoleic acid bleaching assay from leaf and fruit extract

5.5.3 Hydrogen Peroxide Scavenging Activity

Fruit extracts have shown higher hydrogen peroxide (H_2O_2) scavenging activity in comparison to leaf extracts, and it may be due to higher saponin content in fruits. In fruit, highest H_2O_2 scavenging activity was observed in methanolic extract followed by ethanolic and aqueous extract (Fig. 5.6). In leaf extract, H_2O_2 scavenging activity percentage decreased from ethanolic to methanolic and aqueous extracts. A number

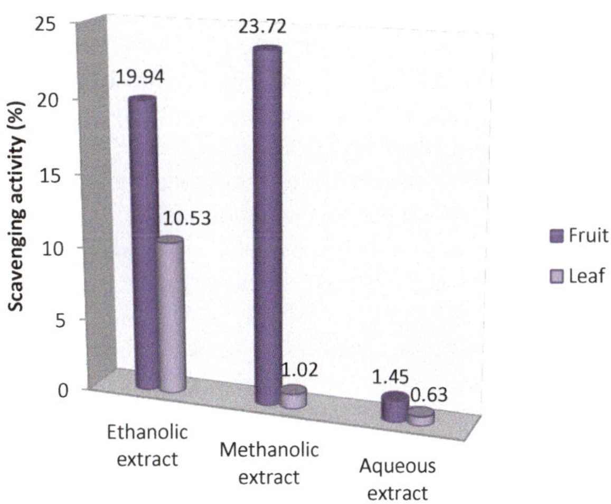

Fig. 5.6 Hydrogen peroxide scavenging activity from fruit and leaf extracts

of workers have reported the H_2O_2 scavenging activity from plant extracts (Yen and Chen 1995; Pal et al. 2011; Keser et al. 2012). In *Crataegus monogyna* Ethanolic extract of *Crataegus monogyna* leaf (100 μg ml^{-1}) showed 30.13% H_2O_2 scavenging activity (Keser et al. 2012).

5.5.4 Reducing Potential from Leaf and Fruit's Extracts

In this assay, the yellow colour of the test solution changes into yellow-green, green and blue after addition of the extracts. The change in colour of test solution indicates the reducing power of extract. The reducing power of all the extracts (leaf and fruits) was increased from lower to higher concentrations in concentration-dependent manner. Maximum reducing power was observed in the aqueous extract of both leaf and fruit (Fig. 5.7). Absorbance is directly related to the reducing potential; it means high absorbance showed high reduction potential of the PE. Aqueous extract of fruit showed more absorbance than aqueous extract of leaf; it means fruit extract has more reducing potential than leaf extracts. Higher absorbance of the reaction mixture indicated upsurge of reducing potential. These reducers show their antioxidant action by interrupting the free radical chain by contributing a hydrogen atom (Gordon 1990).

5.5.5 Estimation of Lipid Peroxidation (LPO) Inhibition Activity

Lipid peroxidation is a free radical inducing oxidative chain reaction in which lipid peroxide was produced as a result of oxidation of lipid molecules and finally causes cell damage. Lipid peroxidation is a substrate-based reaction, i.e. reaction gets stopped after depleting the substrates. All leaf and fruit extracts possess LPO inhibition activity (Fig. 5.8). Methanolic extract of fruit has shown highest LPO inhibition ($IC_{50} = 712$ μg ml^{-1}) than ethanolic ($IC_{50} = 936$ μg ml^{-1}) and aqueous extract ($IC_{50} = 4335$ μg ml^{-1}). In leaf, maximum LPO inhibition activity ($IC_{50} = 802$ μg ml^{-1}) was found in methanolic extract followed by aqueous extract ($IC_{50} = 1897$ μg ml^{-1}) and ethanolic extract ($IC_{50} = 7557$ μg ml^{-1}). Lipid peroxidation inhibition activity is a concentration-dependent activity, and it increased up to an optimum range by increasing the concentration of PE.

For LPO inhibition assay, either enzymatic (Fe/NADPH) or non-enzymatic (Fe/ascorbic acid) method can be used. Non-enzymatic method was performed using yolk as a lipid-rich substrate, so extracts of *S. mukorossi* were found active against non-enzymatic oxidation. Methanolic extract showed higher activity than ethanolic and aqueous extract against lipid peroxidation in both leaf and fruit. Lipid peroxidation inhibition potential of the extracts was directly associated with the antioxidant activity. Non-enzymatic method of lipid peroxidation has also been evaluated in *Pueraria tuberosa* Linn. (Pandey et al. 2007). Phenolic compounds are associated with antioxidant activity, and they also showed inhibition of lipid peroxidation process (Gulcin et al. 2005).

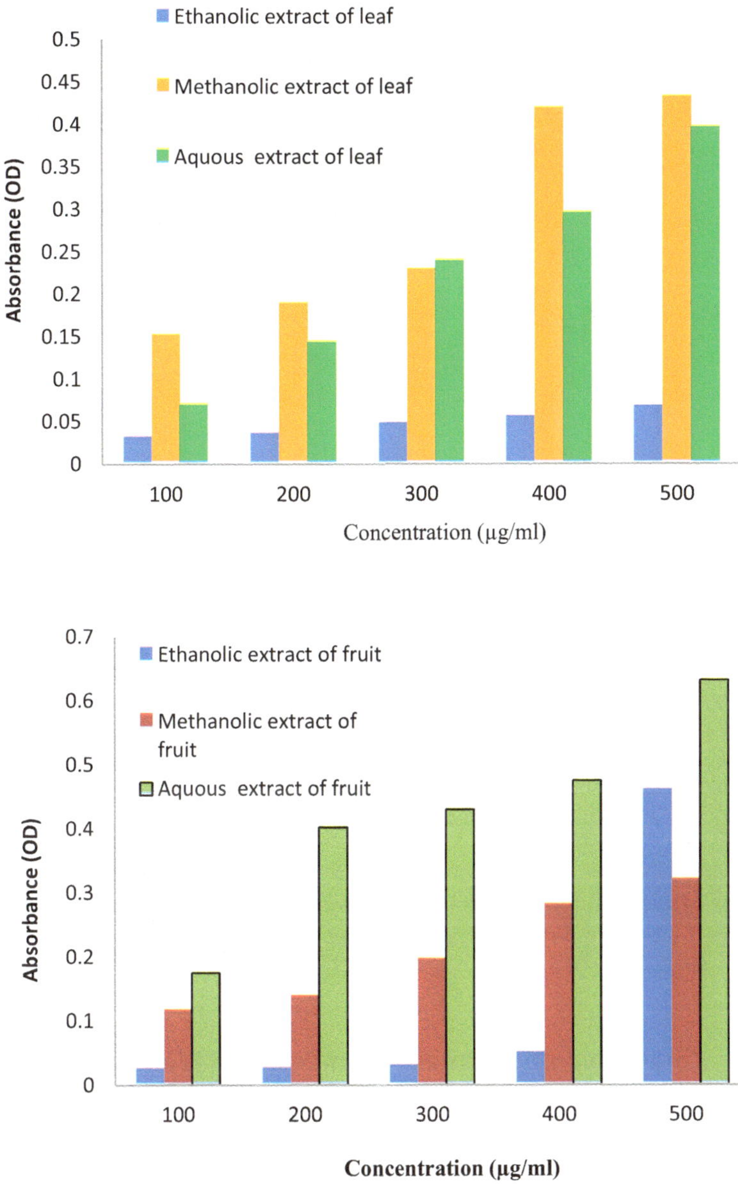

Fig. 5.7 Reducing potential of leaf and fruit extracts. (**a**) Leaf. (**b**) Fruit

5.5.6 Total Phenolic Content from Leaf and Fruit Extracts

Significant amount of phenolics were present in both plant extracts. The results were represented as GAE in reference with standard curve ($y = 0.008x + 0.023$, $R^2 = 0.994$), and TPC value was represented as mg g^{-1} of GAE. In fruits,

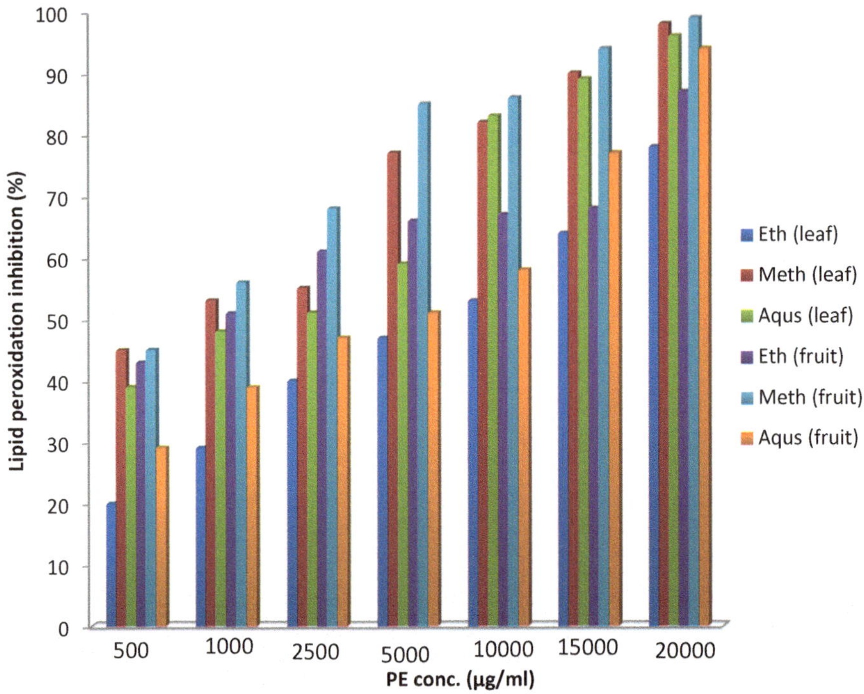

Fig. 5.8 Lipid peroxidation activity of fruit and leaf extracts (IC_{50} was 7557 (eth, leaf), 802 (meth, leaf), 1897 (aqus, leaf), 936 (eth, fruit), 712 (meth, fruit) and 4335 (aqus, fruit). IC_{50} of ascorbic acid is 456.76. All data are highly significant at $p \leq 0.001$)

maximum TPC was recorded in methanolic extract followed by ethanolic and aqueous extracts (Table 5.2). In leaf maximum TPC was found in aqueous extract followed by ethanolic and methanolic extracts.

5.5.7 Total Flavonoid Content from Leaf and Fruit Extracts

Total flavonoid content was reported as rutin equivalent in reference to standard curve ($y = 0.001x–0.065$, $R^2 = 0.992$). Highest TFC was found in methanolic extracts than ethanolic and aqueous extracts of leaf. Whereas in fruit, highest TFC was recorded in aqueous extract when compared to ethanolic and methanolic extracts (Table. 5.4). Flavonoids play a vital role in protection against variety of disease conditions and against human diseases like lipid peroxidation involved in hepatotoxicity, carcinogenesis, atherogenesis, thrombosis, etc. (Young and Wood 2001). Flavonoids also serve as health-promoting compound due to its chemical nature and actions (Havsteen 1983).

Table 5.4 Quantification of TPC and TFC in various extracts of leaf and fruits

Plant material	Extract	TPC (mg g^{-1})	TFC (mg g^{-1})
Leaf	Ethanolic	95.45 ± 0.52	474.03 ± 0.48
	Methanolic	205.16 ± 0.44	540.11 ± 0.89
	Aqueous	337.39 ± 0.26	339.26 ± 0.52
Fruit	Ethanolic	95.52 ± 0.50	525.77 ± 0.23
	Methanolic	469.00 ± 0.57	417.67 ± 0.41
	Aqueous	41.37 ± 0.25	25.20 ± 0.11

5.6 Phytochemical Analysis from In Vitro Cultures of *S. mukorossi*

Most of the medicinal plants possess antioxidant potential and serve as good source of natural antioxidant. Therefore pharmaceutical and scientific communities are concentrating on production of potential pharmaceuticals and natural antioxidants. Sometimes the medicinal plants fall under the category of threatened or endangered due to overexploitation of naturally grown of cultivated plants. In vitro culture can be an alternative source for the production of phytochemicals or metabolites without harming the natural population of the medicinal plants. Plant tissue culture makes the rapid multiplication of selected genotypes possible, allowing the useful metabolites to be collected in greater quantities (George 2008).

In vitro cultures and plants are rich source of secondary metabolites. Production of secondary metabolites from in vitro cultures has a long history (Verpoorte et al. 2002). In vitro cultured plant cells synthesize, accumulate and sometimes exude different classes of secondary metabolites (Matkowski 2008). Production of alkaloids, saponins, polyphenols, terpenes, anthraquinones, cadenolides, etc. have been reported from in vitro cultures and reviewed by several authors (Vanisree and Tsay 2004; Matkowski 2008). Although, in vitro regeneration method is costly, it offers several advantages such as simpler extraction and purification from interfering matrices, independence from climatic factors and seasons, avoiding loss of biodiversity and more control over biosynthetic routes for production of more desired metabolites (Matkowski 2008). Some novel products have been found in in vitro culture, and quantity of compound can be increased or modified by altering the culture conditions. Various natural compounds and antioxidants have been produced from different in vitro plant materials such as leaf, callus, suspension culture, etc. in many plants (Table 5.5).

5.6.1 Collection of In Vitro Materials and Preparation of Extracts

Callus induction and plant regeneration were done by following the method of Singh et al. (2015). Four-week-old callus was collected and dried at 40–45 °C in hot air oven for 4–6 h. Dried callus was used to make coarse powder. Remaining process of

Table 5.5 Various in vitro culture systems were used for production of antioxidants

Culture system	Plant name	Antioxidants methods	Compounds present	References
Callus	*Crocus sativus*	Alleviation of oxidative stress in cultured mammalian cells	Crocin	Ochiai et al. (2004)
	Daucus carota	Lipid peroxidation	Anthocyanins	Ravindra and Narayan (2003)
	Cynara cardunculus	TBRAS	Cynarin, chlorogenic acid	Trajtemberg et al. (2006)
	Hemidesmus indicus	Many in vitro chemical assays	Rutin	Ravishankara et al. (2002)
	Rosmarinus officinalis	Oxidative stress reduction in living cells and many chemical in vitro assays	Carnosic acid	Wijeratne and Cuppett (2007)
	Stevia rebaudiana	FRAP, DPPH	Flavonoids	Tadhani et al. (2007)
Cell suspension	*Anchusa officinalis*	Many in vitro chemical assays	Rosmarinic acid	Petersen and Simmonds (2003) and Soobrattee et al. (2005)
	Beta vulgaris	DPPH	Betalains	Pavlov et al. (2005, 2006)
	Cistanche deserticola	DPPH	Phenylethanoid glycosides	Cheng et al. (2005)
	Daucus carota	Lipid peroxidation	Anthocyanins	Ravindra and Narayan (2003)
	Petroselinum sativum	In vivo (rats)	Flavonols, flavones	Hempel et al. (1999)
	Torreya nucifera	LDL oxidation, nitric oxide inhibition	Abietanediterpenoids	Lee et al. (2006)
	Vitis vinifera	DPPH, lipid peroxidation	Stilbenes, procyanidins	Fauconneau et al. (1997)

Callus and cell suspensions, bioreactor	*Lavandula officinalis*	Superoxide radical scavenging	Rosmarinic acid	Kovacheva et al. (2006)
Flower cell culture	*Ajuga reptans*	β-carotene bleaching, lipid peroxidation	Anthocyanins	Terahara et al. (2001)
	Scutellaria involucrata	Many chemical in vitro assays	Baicalin, wogonoside	Huang et al. (2006)
Hairy roots	*Beta vulgaris*	DPPH	Betalains	Pavlov et al. (2005, 2006)
	Withania somnifera	DPPH, β-carotene/linoleic acid oxidation, brain lipid peroxidation, OH^- radical scavenging	Withanolides	Kumar et al. (2005)
Hairy roots overexpressing CHI under camv35s	*Saussurea involucrata*	H_2O_2-induced cell damage	Apigenin	Fan and Yue (2003)
Shoot culture in bioreactor	*Artemisia judaica*	DPPH	Flavonoids	Liu et al. (2004)
Shoot culture	*Hemidesmus indicus*	Many in vitro chemical assays	Rutin	Ravishankara et al. (2002)
	Rosmarinus officinalis	Oxidative stress reduction in living cells and many chemical in vitro assays	Carnosic acid	Wijeratne and Cuppett (2007)

Table 5.6 Methods used for estimation of different phytochemical activities

Phytochemical/compound	method	Standard used	References
Antioxidant potential	TLC	NA	Tepe et al. (2005)
	DPPH assay	Ascorbic acid	Yu et al. (2008)
	β-carotene/linoleic acid bleaching assay	Ascorbic acid	Miller (1971)
Hydrogen peroxide scavenging capacity	Spectroscopic method	NA	Ruch et al. (1989)
Reducing potential	Spectroscopic method	Ascorbic acid	Nabavi et al. (2008)
Lipid peroxidation inhibition	TBARS	Ascorbic acid	Ohkawa et al. (1979)
Total phenolic content	Folin-Ciocalteu assay	Gallic acid	McDonald et al. (2001)
Total flavonoid content	Spectroscopic method	Rutin	Zhu et al. (2010)

extraction was similar to the leaf and fruit extraction. In vitro leaves were collected from micropropagated plantlets. Leaves (200 mg) were grounded in 10 ml of solvents using chilled mortar and pestle. Extract was filtered, and final volume of extract was maintained up to 20 ml and stored at −20 °C till use. All the standard protocols were used as mentioned in Table 5.6.

5.6.2 Free Radical Scavenging Activity

DPPH-based antioxidant assay is easier and commonly preferred for measurement of antioxidant activity of plant sample. This assay is based on scavenging of DPPH free radical by addition of antioxidants (plant extract), and after the neutralization of free radicals, purple colour of DPPH solution converts to light green to pale yellow (Krishnaiah et al. 2011). Callus and IVL showed significant antioxidant activity (Fig. 5.9). Methanolic extracts have shown higher antioxidant activity in comparison to ethanolic and aqueous extracts of both callus and IVL. In IVL methanolic extracts have higher scavenging activity (IC_{50} = 602.28 µlml^{-1}) than ethanolic (IC_{50} = 741.62 µlml^{-1}) and aqueous extracts (IC_{50} = 1081.93 µlml^{-1}). In callus, highest antioxidant activity was observed in the methanolic extract (IC_{50} = 605.52 µg ml^{-1}) when compared to ethanolic (IC_{50} = 633.27 µg ml^{-1}) and aqueous extract (IC_{50} = 743.54 µg ml^{-1}). In vitro cultures such as leaf, callus, cell suspension cultures and shoots of various plants have also been reported for the presence of antioxidant activity (Lugato et al. 2014; Liu et al. 2004; Vogelsang et al. 2006; Trajtemberg et al. 2006).

In *Stevia* antioxidant activities of callus extract were significantly higher than leaf extracts (Tadhani et al. 2007). In *Habenaria edgeworthii*, antioxidant activity in

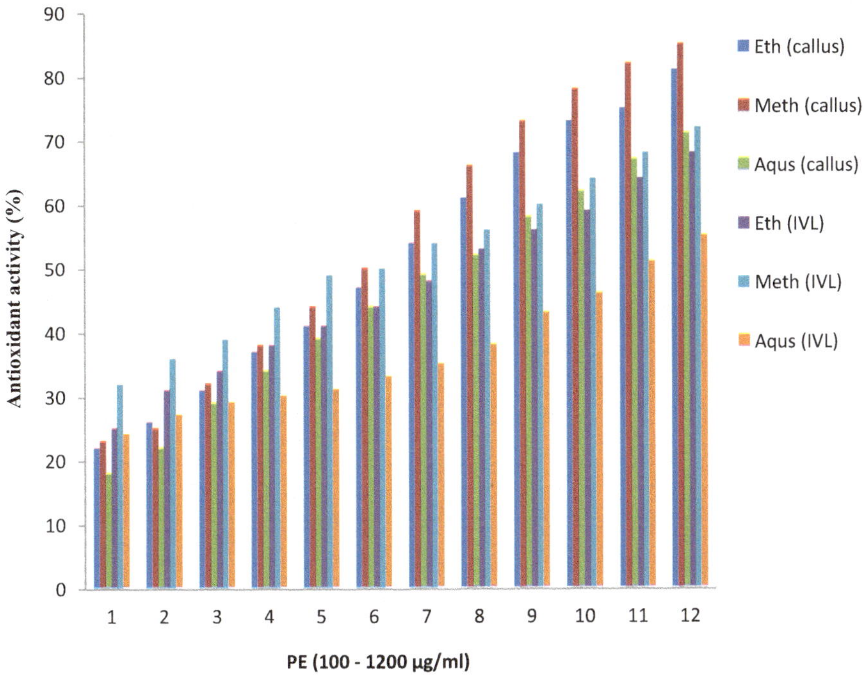

Fig. 5.9 Antioxidant activity of callus and in vitro leaf extracts (IC_{50} was recorded 633 (eth), meth (605) and 743 (aqus) extract of callus. IC_{50} was recorded 741 (eth), 602 (meth) and 1081 (aqus) extract of IVL. IC_{50} of ascorbic acid was 8.34. Concentration of sample for callus (μg ml^{-1}) and in vitro leaf (μl ml^{-1}). All the data is significant at $p \leq 0.001$)

callus culture was significantly higher than that of wild tubers (Giri et al. 2012). In our study, high inhibition of free radicals was observed in methanolic extract of IVL than methanolic extract of callus. Similar result has been reported in *Passiflora alata,* lower antioxidant activity of callus extract (Lugato et al. 2014). In *S. mukorossi,* lower antioxidant potential was observed in callus extracts than in vivo material (leaf and fruit) extracts. Similar to our results, lower antioxidant activity of callus extract has been also reported by other researchers (Hegazi 2011; Sharma and Patni 2013).

5.6.3 Antioxidant Activity Through β-Carotene/Linoleic Acid Bleaching Assay

β-carotene/linoleic acid method is used to estimate the antioxidant potential to inhibit lipid peroxidation (Lai and Lim 2011). β-carotene characteristically produces orange colour, and rate of bleaching of reaction mixture solution was measured by the difference between the initial absorbance and final absorbance reading. PE inhibited the oxidation of linoleic acid. Both callus and IVL extracts showed the presence of

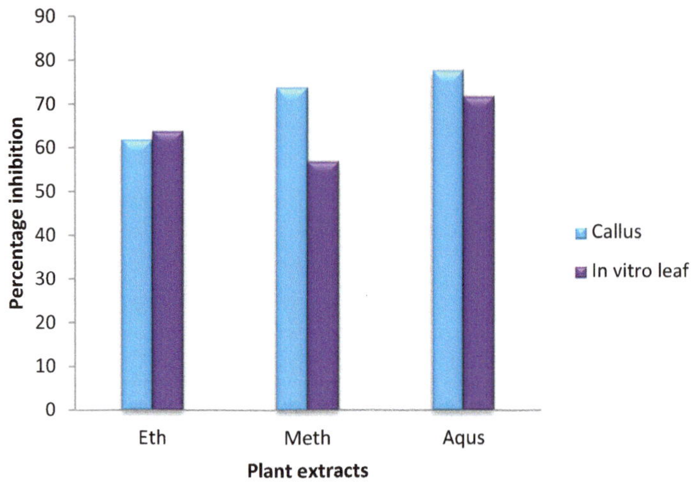

Fig. 5.10 β-carotene/linoleic bleaching assay-based antioxidant activity

significant antioxidant activity. In callus extract, aqueous extract showed highest percentage inhibition (78.44%) than ethanolic (62.57%) and methanolic (74.40%). In IVL, highest percentage inhibition was shown in methanolic extract (72.84%), whereas percentage inhibition in aqueous (57.20%) and ethanolic (64.95%) was comparatively low (Fig. 5.10). In β-carotene/linoleic acid assay, maximum antioxidant activity was present in aqueous extract of callus, while in IVL maximum inhibition percentage was recorded in methanolic extract. Maisarah et al. (2013) have also assessed antioxidant activity through β-carotene bleaching assay, and they found presence of significant antioxidant activity in methanolic extract of different parts of *Carica papaya.* Kumar et al. (2015) also assessed antioxidant activities through various methods from callus extracts in *Pelargonium sidoides*, and they have used TDZ and piclogram for optimization of maximum antioxidant activities.

5.6.4 Reducing Potential (RP)

Reducing potential of the PE was observed on the basis colour of test solution changed from yellow to yellow-green and green to blue. Absorbance of test solution increased from lower to higher concentration of extract, which indicates increase in RP. All extracts of callus and IVL have shown significant level of reducing potential for oxidants present in test solution. Reducing potential of methanolic extract was higher in comparison to ethanolic and aqueous extract of both plant materials (Fig. 5.11). The RP of all the extracts of callus and IVL increased in concentration-dependent manner from lower to higher concentrations. Increase in

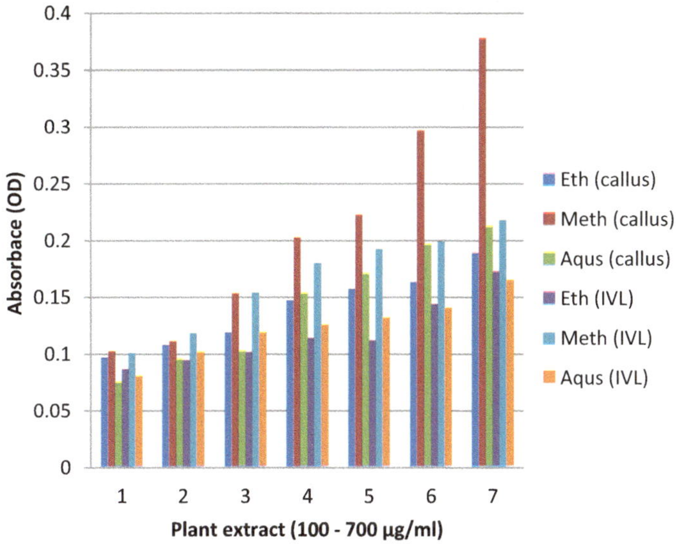

Fig. 5.11 Reducing potential of callus and IVL extracts

RP means increase in antioxidant activity. In this assay, reactions between sample extract (antioxidant compounds) and potassium ferricyanide, trichloroacetic acid and ferric chloride take place and finally form coloured complex, having characteristic absorbance at 700 nm (Alam et al. 2013).

5.6.5 Lipid Peroxidation (LPO) Inhibition Activity

During the physiological processes, lipid peroxidation is an autocatalytic process which may be the reason of cell death (Alam et al. 2013). In in vitro condition, LPO process was induced by ferrous sulphate in lipid-rich egg yolk homogenates; all the extracts of callus and IVL significantly showed the inhibition of LPO in a concentration-dependent manner. Callus extracts showed lower LPO inhibition activity than IVL extracts (Fig. 5.12). In IVL highest LPO inhibition activity was observed in methanolic extract ($IC_{50} = 11.70$ µl ml^{-1}) in comparison to ethanolic ($IC_{50} = 44.31$ µl ml^{-1}) and aqueous extract ($IC_{50} = 46.24$ µl ml^{-1}). In callus, ethanolic extract showed higher activity with lowest IC_{50} value (764.13 µg ml^{-1}), when compared to methanolic ($IC_{50} = 1333.54$ µg ml^{-1}) and aqueous extract ($IC_{50} = 4413.33$ µg ml^{-1}). Lipid peroxidation activity of callus extract was very low in comparison to the other activities such as antioxidant activity, TPC, TFC and reducing potential. Similarly, lower LPO inhibition activity was also observed in callus extract of *Phyllanthus fraternus* (Upadhayay et al. 2013).

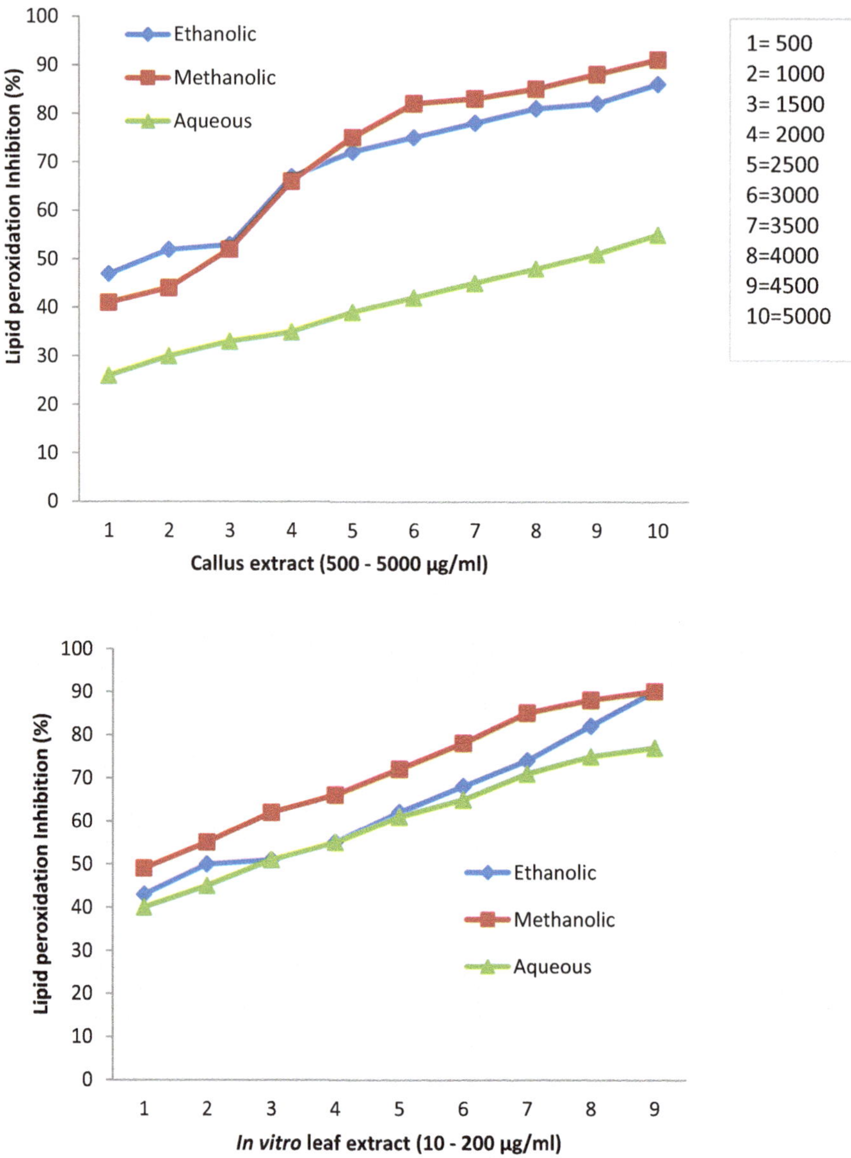

Fig. 5.12 Lipid peroxidation inhibition activity of callus and IVL extracts. (**a**) Callus extracts (IC_{50} was 764 (eth,), 1333 (meth), 4413 (aqus) and 421 (ascorbic acid). All data are significant at $p \leq 0.001$. (**b**) LPO inhibition activity of IVL extracts (IC_{50} was 44 (eth,), 12 (meth), 46 (aqus) and 421 (ascorbic acid). All data are significant at $p \leq 0.001$)

5.6.6 Total Phenolic Content

Total phenolic content was reported as gallic acid equivalents (GAE) in reference to standard curve ($y = 0.008x + 0.031$, $R^2 = 0.996$). In IVL, TPC was highest in ethanolic when compared to methanolic and aqueous extract. Ethanolic extract of callus showed highest TPC than methanolic and aqueous extract (Table 5.6) (Singh 2015). Higher TPC was found in IVL than callus extract. In both callus and IVL extract, maximum TPC was found in ethanolic extract and minimum in aqueous extract. It may be due to high solubility of phenolic compounds in ethanol. Present study indicates a direct relation of TPC with antioxidant activity. In IVL extract, higher TPC may be due to the presence of chlorophyll, while chlorophyll was either absent or present in negligible amount in callus. Similarly, low phenolic content in callus extract was reported by Lugato et al. (2014). Lower TPC and antioxidant activity of callus extract were also found in wild plants of three species of *Ephedra* (Parsaeimehr et al. 2010).

5.6.7 Total Flavonoid Content

Total flavonoid content was reported as rutin equivalent (RE) in reference to standard curve ($y = 0.001x + 0.003$, $R^2 = 0.999$). In IVL, highest TFC was observed in methanolic extract (372.51 ± 0.78 mg g^{-1}) followed by ethanolic (339.12 ± 0.65 mg g^{-1}) and aqueous extract (272.28 ± 1.06 mg g^{-1}) of RE. Similarly in callus, maximum TFC was found (359.22 ± 0.53 mg g^{-1}) in methanolic extract, whereas (315.20 ± 0.70 mg g^{-1}) and (251.16 ± 0.48 mg g^{-1}) of RE were found in ethanolic and aqueous extract, respectively (Table 5.6).

Maximum amount of TFC was observed in methanolic extract of both callus and IVL. Owing to higher TPC and TFC, IVL extract was found as better material than callus for the production of natural antioxidants. The production of polyphenolic compounds especially flavonoids is a light-regulated process; therefore, material with high chlorophyll content showed high TFC. Similar finding was reported by Kim et al. (2011) in *Stevia rebaudiana*, and they also found higher TFC in leaf extract compared to callus extract (Table 5.7).

Table 5.7 TPC and TFC in various extracts of callus and in vitro leaf of *S. mukorossi*

Plant material	Extract	Total phenolic content (mg g^{-1})	Total flavonoid content (mg g^{-1})
In vitro leaf	Ethanolic	91.02 ± 0.09	339.12 ± 0.65
	Methanolic	87.68 ± 0.32	372.51 ± 0.78
	Aqueous	14.15 ± 0.54	272.28 ± 1.06
Callus	Ethanolic	85.34 ± 0.52	315.20 ± 0.70
	Methanolic	74.16 ± 0.22	359.22 ± 0.53
	Aqueous	8.16 ± 0.19	251.16 ± 0.48

Questions

1. Describe the role of natural antioxidants.
2. Describe the saponins and its chemical nature.
3. Describe the biosynthesis of saponins.
4. Explain the role of DPPH in an antioxidant assay.
5. Why are the scientists focusing on the production of natural antioxidants?
6. What is the relation between antioxidant potential and reducing potential?
7. Describe the lipid peroxidation and how it is harmful for the cells and tissues.

Further Readings

Augustin, J. M., Kuzina, V., Anderson, S. B., & Bak, S. (2011). Molecular activities, biosynthesis an evolution of triterpenoid saponins. *Phytochemistry, 72*, 435–457.

Singh, R. (2015). *Somatic embryogenesis and phytochemical analysis of Sapindus mukorossi Gaertn.* PhD thesis, Banaras Hindu University, Varanasi, UP, India.

Singh, R., & Kumari, N. (2015). Comparative determination of phytochemicals and antioxidant activity from leaf and fruit of *Sapindus mukorossi* Gaertn. – A valuable medicinal tree. *Industrial Crops and Products, 73*, 1–8.

Pharmacological Activities and Medicinal Implications of *Sapindus* spp.

6

Indian traditional medicine has a long history of therapeutic applications of herbal drugs. Herbal medicine has been extensively used in traditional medicine system for primary health care and as supplements for promoting the health. Green plants synthesize and preserve a variety of biochemical products, many of which are extractable and used as chemical feedstocks or as raw material for various scientific investigations. These secondary metabolites are responsible for the various pharmacological activities. Many secondary metabolites of plant are commercially important and find use in a number of pharmaceutical compounds. Plants, especially used in Ayurveda, can provide biologically active molecules and lead structures for the development of modified derivatives with enhanced activity and/or reduced toxicity. The small fraction of flowering plants that have so far been investigated have yielded about 120 therapeutic agents of known structure from about 90 species of plants. Some of the useful plant drugs include vinblastine, vincristine, taxol, podophyllotoxin, camptothecin, digitoxigenin, gitoxigenin, digoxigenin, tubocurarine, morphine, codeine, aspirin, atropine, pilocarpine, capsaicin, allicin, curcumin, artemisinin and ephedrine, among others. In some cases, the crude extract of medicinal plants may be used as medicaments. On the other hand, the isolation and identification of the active principles and elucidation of the mechanism of action of a drug are of paramount importance. Hence, works in both mixture of traditional medicine and single active compounds are very important. Where the active molecule cannot be synthesized economically, the product must be obtained from the cultivation of plant material. Only a small percentage of medicinal plants traded in India are solely cultivated. The obligatory demand for a huge raw material of medicinal plants is met from their wild populations. Over-exploitation and destructive harvesting to meet such demand in fact threaten the survival of many rare species (Natesh 1999). Confronted by such unprecedented genetic erosion and disappearance of species and ecosystems, conservation of natural resources assumes paramount urgency.

R. Singh, B. Sharma, *Biotechnological Advances, Phytochemical Analysis and Ethnomedical Implications of Sapindus species*,
https://doi.org/10.1007/978-981-32-9189-8_6

Sapindus species also contain the variety of phytochemicals, and these phytochemicals are responsible for the different biological activities. Phytochemical studies of ethanolic and aqueous extract of fruits of *Sapindus* showed the presence of alkaloids, phytosterols, phenolic compounds, tannins, flavonoids, glycosides and saponins (George and Shanmugam 2014). Kirtikar and Basu (1991) reported that the fruits of this plant were used for the treatment of various diseases like excessive salivation, pimples, epilepsy, chlorosis, migraines, eczema and psoriasis. Several authors reported various biological activities from different parts of this plant such as leaf, fruits pericarp, seeds, whole fruits, etc.

6.1 Biological Activities of *Sapindus mukorossi*

Genus *Sapindus* is a popular economical and medicinal tree and also documented in the ancient medicine literature. Almost every part of this plant has been reported for their biological activities. *S. mukorossi* is most explored species of *Sapindus* for the phytochemistry and pharmacological studies. Few other species named *S. trifoliatus*, *S. saponaria*, *S. emarginatus*, etc. have been documented for their medicinal values, and now scientists started the validation of these medicinal values. All the medicinal properties of *S. mukorossi* are summarized in Table 6.1.

6.1.1 Antioxidant Activity

Oxidation and antioxidation (reduction) are natural phenomena in a human body. Inside the human body, free radical species such as reactive oxygen species (ROS), reactive nitrogen species (RNS), etc. are continuously generated from the number of metabolisms. Several antioxidant enzymes are also naturally present in body and acting against free radicals. Antioxidants are a molecules, chemicals or compounds that neutralize the free radicals and protect from the damage caused by the free radicals. Natural antioxidants are more safe and cost-effective than synthetic ones (Singh and Kumari 2015). Ethanolic, methanolic and aqueous extracts of leaf, fruit, callus and in vitro leaf of *S. mukorossi* were used for antioxidant activity estimation (Singh and Kumari 2015). Antioxidant activity was evaluated by several antioxidant assays, namely, DPPH assay, β-carotene/linoleic acid bleaching assay, reducing power assay, etc. Significant antioxidant activity of extracts has been recorded through the different assay (Singh and Kumari 2015; Singh 2015). Of these extracts aqueous extracts of fruit have shown the highest antioxidant potential.

6.1.2 Antimicrobial Activity

Most of the microbes such as bacteria, fungi, viruses, etc. are pathogenic in nature and caused infectious diseases including skin and gonorrhoeal diseases. Antimicrobial activity of *S. mukorossi* against common dental caries pathogen was reported by

Table 6.1 Pharmacological activities of *Sapindus mukorossi*

Biological activities	Plant parts used for	Extracts/ compounds	Microorganisms/models/ details	References
Antibacterial	Leaf and fruit	Eth, meth and aqus	*S. aureus*, *E. faecalis*, *E. aerogenes*, *S.* Typhimurium, *E.coli*, *K. pneumoniae*	Singh et al. (2016a, b, c)
	Fruit	Ace, meth, eth, hot aqus and cold aqus	*S. mutans*, *S. aureus*, *L. acidophilus*	Aneja et al. (2010)
	Seed	Eth, meth and hex	*Acinetobacter baumannii* ATCC 19601	Chen et al. (2010)
	Fruit's pericarp	Eth	*E. faecalis*, *S. aureus*, *B. subtilis*, *P. aeruginosa*, *K. pneumoniae*, *B. cereus*, *P. alcaligenes*, *M. luteus*,	Sharma et al. (2013)
		Eth and aqus	*S. aureus*, *E. coli*	Gerorge and Shanmugan (2014)
		Eth and chl	*H. pylori*	Ibrahim et al. (2006)
	In vitro leaf and callus	Eth, meth and aqus	*S. aureus*, *E. faecalis*, *E. aerogenes*, *S.* Typhimurium, *K. pneumoniae*, *E.coli*	Singh et al. (2016a, b, c)
Antifungal	Fruit pericarp	Eth and aqus	*A. niger*, *A. fumigatus*	Gerorge and Shanmugan (2014)
	Fruit	Ace, meth, eth, hot aqus and cold aqus	*C. albicans*, *S. cerevisiae*	Aneja et al. (2010)
	Leaf and fruit	Eth, meth and aqus	*C. albicans*, *C. tropicalis*, *C. parapsilosis*	Singh et al. (2016a, b, c)
	IVL and callus	Eth, meth and aqus	*C. albicans*, *C. tropicalis*, *C. parapsilosis*	Singh et al. (2016a, b, c)
	Pulp	Triterpenoid saponins	*Trichophyton rubrum*, *Candida albicans*	Hu et al. (2018)
Anti-gonorrhoeal	Leaf	Eth, meth and aqus	*N. gonorrhoeae*	Bhargava et al. (2012)
Antiasthmatic	Fruit	Aqus	Using trachea of goat	Parganiha et al. (2011)
Anticancer and cytotoxic	Seed	Meth, eth and hex	Cell lines of human melanoma, lung, liver, prostrate, cervical, osteosarcoma, breast and bladder cancer	Chen et al. (2010)
Antidandruff activity	Seed coat	Eth	*Trichophyton rubrum*	Kaladhar et al. (2013)
Anti-hyper glycaemic and	Fruit	Eth	Albino Wistar rats	Verma et al. (2012)

(continued)

Table 6.1 (continued)

Biological activities	Plant parts used for	Extracts/ compounds	Microorganisms/models/ details	References
anti-hyperlipidaemic				
Anti-inflammatory	NA	Crude saponin and hederagenin	Inhibited the development of carrageen-induced oedema in the rat hind paw as well as on granuloma and exudates formation induced by croton oil in rats	Takagi et al. (1980)
Anti-platelet aggregation	Galls	Extract		Huang et al. (2007)
Anti-*Trichomonas*	NA	Saponin mixture	*Trichomonas vaginalis*	Tiwari et al. (2008)
Anxiolytic	Whole fruits with seeds	Meth		Chakraborty et al. (2010)
Hepatoprotective	Fruit pericarp		Primary hepatocyte culture and rat model of CCl_4-mediated liver injury	Ibrahim et al. (2008)
Repellency and insecticidal			*Sitophilus oryzae* and *Pediculus humanus*	Rahman et al. (2007)
Molluscicidal	NA	Extract	*Pomacea canaliculata*	Huang et al. (2003)
	Fruit pericarp	Chl, ether, carbon tetrachloride, ace and eth	*Lymnaea acuminate*	Upadhyay and Singh (2011)
Piscicidal	Pericarp		*Heteropneustes fossilis* and *Channa punctuate*	Virdi (1982)
Spermicidal activity	Fruits pericarp		Rabbits	Pelegrini et al. (2008)
Tyrosinase inhibition activity	Seed	Meth, eth and hex	In vitro mushroom tyrosinase assay	Chen et al. (2010)

Eth ethanolic, *Meth* methanolic, *Hex* hexane, *Aqus* aqueous, *Chl* chloroform, *Ace* acetone

List of microbes: *A. baumannii (Acinetobacter baumannii), A. niger (Aspergillus niger), A. fumigatus (Aspergillus fumigatus), A. versicolor (Aspergillus versicolor), A. ochraceus (Aspergillus ochraceus), B. subtilis (Bacillus subtilis), B. cereus (Bacillus cereus), C. albicans (Candida albicans), C. tropicalis (Candida tropicalis), C. parapsilosis (Candida parapsilosis), E. aerogenes (Eenterobacter aerogenes), E. cloacae (Enterobacter cloacae), E. faecalis (Enterococcus faecalis), H. pylori (Helicobacter pylori), K. pneumoniae (Klebsiella pneumoniae), L. acidophilus (Lactobacillus acidophilus), L. monocytogenes (Listeria monocytogenes), M. flavus (Micrococcus flavus), M. luteus (Micrococcus luteus), N. gonorrhoeae (Neisseria gonorrhoeae), P. aeruginosa (Pseudomonas aeruginosa), P. alcaligenes (Pseudomonas alcaligenes), P. funiculosum (Penicillium funiculosum), P. verrucosum (Penicillium verrucosum), P. ochrochloron (Penicillium ochrochloron), S. mutans (Streptococcus mutans), S. aureus (Staphylococcus aureus), S. cerevisiae (Saccharomyces cerevisiae), S.* Typhimurium *(Salmonella* Typhimurium*), T. rubrum (Trichophyton rubrum), T. vaginalis (Trichomonas vaginalis), T. viride (Trichoderma viride)*

Aneja et al. (2010). They used acetonic, methanolic, ethanolic, hot aqueous and cold aqueous fruit extract for conducting this study. They conducted their study against *S. mutans*, *S. aureus*, *L. acidophilus*, *C. albicans* and *S. cerevisiae*. Result showed that the all extract was effective against only *S. cerevisiae*. The largest zone of inhibition (29.65 ± 0.57) is obtained from acetonic extract and smallest (19.65 ± 0.57) with hot aqueous extract.

Antibacterial activity from ethanolic extract of fruits pericarp was studied (Sharma et al. 2013) on some selected bacteria. They carried out the experiments with well diffusion method, and extract was prepared in 20% DMSO. They reported that extract (50 and 100 μg/ml) was effective against all tested microorganisms (*E. faecalis*, *P. aeruginosa*, *S. aureus*, *A. denitrificans*, *K. pneumoniae*, *B. cereus*, *P. alcaligenes*, *M. luteus* and *B. subtilis*). Zone of inhibition was maximum in *E. faecalis* (17 mm) from 50 μg/ml and in *P. aeruginosa* (20 mm) from 100 μg/ml plant extract.

Gerorge and Shanmugan (2014) reported antimicrobial activity of fruits pericarp extract against *S. aureus*, *E. coli*, *Aspergillus niger* and *A. fumigatus*. They used well diffusion method and 25%, 50%, 75% and 100% plant extract for filling the well. Ethanolic extract was more effective than aqueous extract against all tested microorganisms. Among all percentage of plant extract, 100% was most effective. Maximum zone of inhibition (11 mm) was found against *E. coli* with ethanolic extract coli, and minimum (9 mm) was against *S. aureus*. 25% plant extract was ineffective against all tested microorganisms.

Fruit and leaf extracts were tested against clinical pathogens (Singh et al. 2016a, b, c). They tested the efficacy of extract against *S. aureus*, *E. aerogenes*, *E. faecalis*, *S.* Typhimurium, *E. coli*, *K. pneumoniae*, *C. albicans*, *C. tropicalis* and *C. parapsilosis*. Both fruit and leaf extract have shown potential antimicrobial activity against most of the tested pathogens. In fruit extract, methanolic and aqueous extract have shown significant antimicrobial activity against all tested organisms except *C. albicans*. All extract of fruit showed maximum inhibition zone (diameter) against *C. tropicalis*, methanolic extract (12.50 ± 0.29), aqueous extract (11.66 ± 0.33) and ethanolic extract (10.16 ± 0.16). In leaf extract, the highest inhibition zone (9.66 ± 0.52) was observed by ethanolic extract against *S.* Typhimurium. Fruit and leaf extracts have shown more activity against Gram-negative bacteria.

Callus and in vitro leaf extracts were tested against a number of clinical pathogens (*S. aureus*, *E. aerogenes*, *S.* Typhimurium, *V. cholerae*, *E. coli*, *K. pneumoniae*, *P. aeruginosa*, *C. albicans*, *C. tropicalis* and *C. parapsilosis*), and significant result was obtained (Singh et al. 2016a, b, c). Both plant extracts of *S. mukorossi* have shown antimicrobial activity. Methanolic extract of IVL showed maximum efficacy against *C. tropicalis* (8.06 ± 0.17). Ethanolic extracts of IVL showed highest antimicrobial efficacy against *V. cholerae* (8.00 ± 0.06). Ethanolic extracts of callus have shown significant antimicrobial activity against *E. coli* (8.20 ± 0.11). All the materials of *S. mukorossi* have the potential antibacterial and antifungal activity.

6.1.3 Antiasthmatic Activity

Asthma is a very common long-term inflammatory disease of the bronchi of the lungs (Asthma Fact sheet, WHO 2013). Asthma is characterized by variable and recurring symptoms, reversible bronchi obstacle and bronchospasm (NHLBI Guideline 2007). These symptoms include events of coughing, chest tightness, wheezing and shortness of breath. Mainly genetic and environmental factors such as air pollution and allergens are responsible for the cause of asthma (Asthma Fact sheet, WHO 2013; Martinez 2007). Side effects of long-term medication such as aspirin and beta blockers may act as a potential triggers for asthma (Wikipedia 2018).

In vitro antiasthmatic study was performed by Parganiha et al. (2011) by using the trachea of goat. The trachea was cut into individual ring and tied together in series to form a chain. The tracheal chain was suspended in bath of Krebs solution (having various chemicals) for conducting the experiments. The aqueous fruit extract at the dose 380 mcg ml^{-1} and 640 mcg ml^{-1} significantly inhibited acetylcholine-induced bronchoconstriction of isolated goat trachea. They concluded that aqueous extract of fruit has significant antiasthmatic potential.

6.1.4 Anti-gonorrhoeal Activity

The sexually transmitted diseases (STDs) are most common public health problems around the world, and these are considered as prime factor for transmission of HIV/AIDS. Gonorrhoea is one of the common sexually transmitted infections, caused by ***Neisseria gonorrhoeae*** **or** ***gonococcus*** **(bacteria).** The bacteria are mainly found in discharge from the penis and in vaginal fluid and easily transmitted between people through unprotected sex. In Nepal, *S. mukorossi* is commonly used for treatment of gonorrhoeal disease. The leaf extract (ethanolic, methanolic and aqueous) of *S. mukorossi* was used against *N. gonorrhoeae* for screening of antimicrobial activity (Bhargava et al. 2012). They also used well diffusion method for the experiments. They performed their experiments against different strain of *N. gonorrhoeae* that was resistant for various antibiotics. Maximum zone of inhibition was (12.5 ± 0.5) for ethanolic and 9.5 ± 0.4 for methanolic extract. They observed the significant anti-gonorrhoeal activity and concluded that this activity is due to presence of saponin in the extract.

6.1.5 Antidandruff Activity

Nowadays dandruff is a common but very serious hair problem and main reason behind the hair fall. Growth of *Trichophyton rubrum* on scalp is one of the major causes for dandruff. Ethanolic extract of seed coat showed significant activity against *T. rubrum* (Kaladhar et al. 2013). Extract showed zone of inhibition from 20 μl. They also reported that protein interaction studies of MEK_1 and MEK_2 of *C. albicans*, *S. cerevisiae* and *Pichia stipitis* have shown network interaction with other proteins like CDH_3, GLN_3, HEM_{12}, etc. MEK_1 and MEK_2 proteins are also present in *T. rubrum*, so this study would be useful for inhibition of dandruff-causing microbes.

6.1.6 Anti-platelet Aggregation Activity

Two saponins named sapinmusaponins Q and R were isolated from the fractionation of ethanolic extract of *S. mukorossi (*gall's extracts). Sapinmusaponins Q and R showed more potent anti-platelet aggregation activity than standard drug named aspirin (Huang et al. 2007). Sapinmusaponins F-J isolated from the galls of *S. mukorossi* showed anti-platelet-aggregation effects, but no obvious cytotoxic activity for platelets as assayed by lactate dehydrogenase (LDH) leakage was reported (Huang et al. 2006).

6.1.7 Anti-inflammatory Activity

Inflammation word is originated from a Latin word *inflammatio*. Inflammation is part of the complex biological response of body tissues which is result of harmful stimuli such as pathogens, injury or any irritants. Redness, heat, pain and swelling are the main symptoms of inflammation. Takagi and co-workers (1980) investigated the anti-inflammatory activities of hederagenin and crude saponin isolated from *S. mukorossi*, using carrageenan induced oedema, granuloma pouch and adjuvant arthritis in rats. The effects of these agents on vascular permeability and acetic-acid-induced writhing in mice were also examined. Anti-inflammatory activity on carrageenan oedema was observed after intraperitoneal and oral administration of crude saponin, whereas hederagenin and the other agents showed activity only when administered intraperitoneally. They observed that crude saponin showed a significant inhibitory effect on granuloma and exudate formations in rats, inhibited the increase in vascular permeability and the number of writhings induced by acetic acid in mice and significantly inhibited the development of hind-paw oedema associated with adjuvant arthritis in rats after oral administration. They concluded that crude saponin has shown some degree of anti-inflammatory activity.

6.1.8 Anxiolytic Activity

Human anxiety is defined as a feeling of uneasiness, uncertainty or tension stemming from the expectation of imagined or unreal threat. About one-eighth population worldwide is affected by anxiety and has become an important research area in the field of psychopharmacology. Anxiolytic activity has been evaluated using methanolic extracts of whole fruits with seeds (Chakraborty et al. 2010). The anxiolytic activity was evaluated by elevated plus maze, Y-maze, hole board, marble-burying behaviour models and actophotometer using albino mice of either sex. Methanolic extract of fruits and seeds was used for the study, and 200 and

400 mg/kg extracts were compared with diazepam (2 mg/kg) and fluoxetine (10 mg/kg) as the standard anxiolytic drugs. Based on the result observed from the abovementioned experimental model, PE showed the significant anxiolytic activity. The extract significantly increased the number of entries and time spent in the open arm in the elevated plus maze and also increased the number of head dipping and line crossing. The result also showed the decreased numbers of visits to the three arms, locomotor score and number of marble-buried in hole board, Y-maze, actophotometer and marble-burying behaviour models, respectively (Chakraborty et al. 2010). Present study showed that both the doses are effective as anxiolytic agents.

6.1.9 Hepatoprotective Activity

A number of plants are possessing hepatoprotective activity. Ibrahim et al. (2008) studied hepatoprotective effect from extracts of fruit pericarp of *S. mukorossi* (2.5 mg/L) both in vitro on primary hepatocyte cultures and in vivo in a rat model of carbon tetrachloride (CCl_4)-mediated liver injury. *S. mukorossi* extracts showed protective capacity both in vitro on primary hepatocyte cultures by reducing the levels of LDH and GPT that were released from CCl_4-treated rat hepatocytes. The reduction level of LDH and GPT was directly related to the extract concentration. In in vivo model, CCl_4-treated rat serum activities of AST, ALT, ALP and bilirubin were increased significantly when compared to the control. After the treatment with the extract, serum activities of AST, ALT, ALP and bilirubin were decreased, and normal architectural liver pattern was restored. Protective effect of *S. mukorossi* fruit extract in liver cirrhosis was also reported by Ibrahim et al. They used CCl_4 for causing liver cirrhosis.

6.1.10 Insecticidal Activity

S. mukorossi fruit's extracts have the insecticidal potential. Repellency and insecticidal activity of ethanolic fruit extracts were evaluated against *Sitophilus oryzae* and *Pediculus humanus* (Rahman et al. 2007). Fruit extract caused significant mortality and repellency on the target insects, and bioassay results clearly indicated that the toxic and repellent effects have linear correlation with the concentration of extract. Plant extract concentration (3.20%) caused toxicity for *Sitophilus oryzae* (88.1%) and 70.1% for *Pediculus humanus.*

6.1.11 Molluscicidal Activity

Molluscicidal activity has been also evaluated using by some workers. Huang et al. (2003) studied the effect of *S. mukorossi* extract on golden apple snail (*Pomacea canaliculata* Lamarck.) for molluscicidal activity with LC_{50} values of 85, 22 and

17 ppm at 24, 48 and 72 h exposure period, respectively. Potential of fruit pericarp extract of *S. mukorossi* was also evaluated as molluscicides against *Lymnaea acuminata* (Upadhyay and Singh 2011). These snails are the intermediate host of liver fluke *Fasciola gigantica*, which causes 94% fascioliasis in the buffalo population of northern India (Singh and Agarwal 1983).

6.1.12 Piscicidal Activity

Effects of fruit pericarp have been also studied for piscicidal activity. Pericarp was the most toxic part resulting in 100% mortality rate within 12 h, and mean survival time was found to be 1.18 h. Plant extracts possess strong high potential for fish eradication and its LD_{10}, LD_{50} and LD_{100} range from 3.5 10 ppm at 48 h. Fruit pericarp of *S. mukorossi* can be used as a selective piscicidal agent for horny fish like *Heteropneustes fossils* and *Channa punctuate* (Virdi 1982).

6.1.13 Spermicidal Activity

S. mukorossi is the best known species of the genus *Sapindus* and is traditionally used in oriental medicine as a spermicide (Pelegrini et al. 2008). Saponins from *S. mukorossi* are known to be spermicidal. Morphological changes in human ejaculated spermatozoa after exposure to this saponin were evaluated under scanning electron microscopy. The minimum effective concentration (0.05% in spot test) did not affect the surface topography after exposure for 1 min. However, incubation of spermatozoa for 10 min resulted in extensive vesiculation and a disruption of the plasma membrane in the head region. Higher concentrations (0.1%, 1.25%, 2.5% and 5.0%) caused more or less similar changes which included vesiculation, vacuolation, disruption or erosion of membranes in the head region. These findings suggest that the morphological changes observed are due to alterations in the glycoproteins associated with the lipid bilayer of the plasma membrane of spermatozoa. This spermicidal property has been used in contraceptive cream. A contraceptive cream developed by the CDRI (Lucknow) is going to hit the Indian markets. It is advocated to be totally safe and easy to use. It is intended for post-coital use.

Raghuvanshi et al. (2001) developed a spermicide compound, called Praneem polyherbal, featuring antimicrobial traits, from the pericarp of fruit of *S. mukorossi*, leaves of *Azadirachta indica* and oil of *Mentha citrata*. The association of these three plants produced a highly powerful spermicide, which was tested on rabbits and on human sperm through in vitro and in vivo studies. When this spermicide was applied to the vagina of female rabbits, pregnancy failed to occur.

6.1.14 Anti-*Trichomonas* Activity

Saponin mixture isolated from *Sapindus* has been tested for anti-*Trichomonas* activity and found significant result (Tiwari et al. 2008). Lower concentrations (0.005%) of saponin showed the complete inhibition of *Trichomonas vaginalis* growth after 24 and 48 h of incubation. MIC against *T. vaginalis* was 6 μM of metronidazole. Inhibition ability of saponins on parasites are good, it adhere to HeLa cells and decreased the proteolytic activity of the parasite's cysteine proteinases was in concentration-dependent manner. This was associated with the decreased expression of adhesin AP65 and membrane-expressed cysteine proteinase TvCP2 genes. Saponins produced no adverse effect on host cells in mitochondrial reduction potential measurement assay. Saponins also reversed the inhibitory mechanisms that were diminished by *Trichomonas* for evading host immunity. Early response of saponins to disrupt actin cytoskeleton in comparison with their effect on the nucleus suggests a membrane-mediated mode of action.

6.1.15 Anti-lipid Peroxidation Activity

In vitro anti-lipid peroxidation activity was evaluated by Singh and Kumari (2015). They used egg yolk as lipid-rich media for lipid peroxidation assay (TBARS assay). Free radicals cause the lipid peroxidation of the cell/tissue. Various extracts (ethanolic, methanolic and aqueous) of leaf, fruits and in vitro materials were used for investigation of anti-lipid peroxidation activity. Anti-lipid peroxidation activity was increased in the concentration-dependent manner. Of these extracts, methanolic extract of fruits showed highest lipid peroxidation inhibition activity.

6.1.16 Antidiabetic and Anti-hyperlipidaemic Activity

Antihyperglycaemic and anti-hyperlipidaemic activity has been reported from the hydroalcoholic extract of *S. mukorossi* fruits (Verma et al. 2012). Streptozotocin was used to induce the diabetic condition in Wistar albino male rats. Extracts were given through oral administration to treat the diabetes. The diabetic rats were treated for 20 days from the extracts with 250 mg/kg and 500 mg/kg and glybenclamide as standard. The daily administration of this extract significantly decreased the blood glucose level and lipid level. This extract decreased the elevated glucose level in diabetic animals.

6.2 Pharmacological Activities of *S. trifoliatus*

A number of pharmacological activities were reported from the different parts of *S. trifoliatus*. Few important activities are summarized in brief in Table 6.2. Here some of the activities are discussed in details.

Table 6.2 Biological activities of *S. trifoliatus and S. emarginatus*

Sapindus species	Biological activities	Plant parts used	Extract/ compounds	Microorganisms/details	References
S. trifoliatus	Antioxidant	Aerial	Eth	In vivo antioxidant enzymes SOD and catalase	Sahoo et al. (2010a, b)
	Anticancer	Fruit	Extracts	Inhibits proliferation of SKBR3 and MDA-MB435 cell lines (human breast cancer)	Rao and Lakshmi (2012)
	Antidiabetic	Aerial	Eth	Alloxan-induced Wistar albino rats were used. Oral glucose tolerance, single-dose and multi-dose study and estimation of glycosylated haemoglobin	Sahoo et al. (2010a, b)
	Anti-inflammatory	Seed	Eth	Inhibition of leucocyte migration	Arul et al. (2004)
		Fruit's pericarp	Aqus	Inhibition of 5-lipoxygenase, cyclooxygenase, leukotriene B4 and nitric oxide synthase	Arulmozhi et al. (2005a, b, c)
	Analgesic	NA	Eth	Reduce diabetic neuropathic pain	Sahoo et al. (2010a, b)
	Anti-ulcerogenic	Leaf	Meth and aqus	Reduction in pylorus ligation-induced ulcers	Kishore et al. (2010)
	Anti-migraine	Pericarp	Aqus	Anti-migraine hyperalgesic potential	Arulmozhi et al. (2005a, b, c)
	Antiepileptic	Pericarp	Aqus	Anticonvulsant activity	Jayasree et al. (2011)
S. emarginatus	CNS	Fruit	Meth	CNS depressant action, muscles relaxant activity, reduction in spontaneous activity, reduction in exploratory behavioural pattern, inhibition of cocaine-induced hyperactivity, potentiate phenobarbitone sodium-induced sleeping time	Srikanth and Veeranjaneyulu (2005) and Chattopadhyay et al. (2003)
	Antidiabetic	Leaf	Alc	Hypoglycaemic effect	Srikanth and Muralidharan (2009a, b)
	Anti-hyperlipidaemic activity	Fruit	Meth	Lower plasma glycerides and cholesterol level	Jeyabalan and Palayan (2009)

Note: *Eth* ethanolic, *Meth* methanolic, *Aqus* aqueous, *Alc* alcoholic

6.2.1 Anticancer Activity

Cancer is the most deadly disease around the globe. Genetic and epigenetic factors are responsible for the cause of cancer. Different fractions of methanolic extract of leaf (*S. trifoliatus)* were used for the evaluation of in vivo anticancer activity against mammary cell proliferation. They assessed by inducing the Ehrlich's Ascites Cell (EAC) in Swiss albino mice (Pradhan 2014).

S. trifoliatus ethanolic leaf extracts showed anticancer activity against few breast cancer cell lines, namely, MDA-MB435 and SKBR3 (Pradhan and Tripathy 2014). They used cell proliferation assay, cell viability assay and cell cycle analysis assay to determine the cytotoxic capacity of plant extract against cancerous cell lines.

6.2.2 Anti-migraine Activity

Fruit's pericarp extract (aqueous) of *S. trifoliatus* has been used for evaluation of pharmacological studies of central nervous system. Swiss albino mice (male) were used as experimental model. The intraperitoneal administration of extract (20 and 100 mg/kg) significantly decreases the spontaneous locomotor activity, while thiopental-induced sleeping time increased at dose of 100 mg/kg. In receptor radioligand binding studies, ST exhibited affinity towards dopaminergic, α-adrenergic and muscarinic receptors. The aqueous extract of this plant may have the phytocompounds having with potential neuroleptic properties (Arulmozhi et al. 2005a, b, c). This plant extract would be useful for the treatment of migraine.

6.2.3 Analgesic Activity

Antinociceptive and analgesic activity was evaluated using aqueous extracts of *S. trifoliatus* (Arulmozhi et al. 2004). They used various pain models such as chemical- and thermal-induced pain and nitroglycerin-induced hyperalgesia for investigation of pain on inflamed tissue. They also used carrageenan for induction of inflamed paw model. Plant extract (20 and 100 mg/kg) showed the significant anti-neurogenic activity and anti-inflammatory activity in formalin induced in mice. A dose of 20 and 100 mg/kg also increases the pain latency in animals during the hot plate and nitroglycerin-induced hyperalgesia test. A number of workers reported that plant extracts of *S. trifoliatus* exhibited significant analgesic activity. The extracts also possess the inhibitory activity on both central and peripheral pain mechanisms. The extract also modulates the NO-mediated sensitive transmission. It is supposed that *S. trifoliatus* extracts inhibit cyclooxygenase (COX) in peripheral tissues, thus interfering the mechanism of transduction of primary pain receptors.

Arulmozhi et al. (2005a, b, c) also investigate the anti-hyperalgesic activity of *S. trifoliatus* extract using in vivo migraine model. They used several experimental models such as hot plate test and acetic acid-induced abdominal constriction test for the assessment of analgesic activity. Significant activity was observed from

10 mg/kg intraperitoneal administration of extracts in the hot plate test and abdominal constriction test.

6.2.4 Anti-inflammation Activity

Inflammation was created by using various chemicals, namely, carrageenan, histamine, serotonin, zymosan, capsaicin, arachidonic acid, oxazolone and orobenzene (Arulmozhi et al. 2005a, b, c). In vitro inflammation inhibitory activity has been assessed against 5-lipoxygenase (5-LO), cyclooxygenase (COX), leukotriene B_4 and nitric oxide synthase. Acute pedal inflammation was introduced by carrageenan, histamine, serotonin and zymosan in rats and mice; plant extract showed significant effect against inflammation. It is supposed that through the modulation 5-LO and COX pathways, extract showed the anti-inflammation activity. Extract also inhibited the ear inflammation after the topical application. However the extract was ineffective against inflammation induced by oxazolone and orobenzene.

6.2.5 Muscle Relaxation Activity

S. trifoliatus aqueous extracts exhibited the muscle relaxation activity (Jayasree et al. 2012). Extract was prepared from the fruits pericarp. Diazepam was used as standard muscle relaxation drugs. Dose of 200 mg/kg showed the maximum skeletal muscle relaxation and locomotor activity.

6.2.6 Antidiabetic Activity

Diabetic mellitus is a metabolic disorder, arises from unhealthy and imbalance life style. Diabetes affects the metabolisms of carbohydrate, protein and fat. Every year many thousands of people are diagnosed with diabetes, and a number are continuously increasing. Oral hypoglycaemic agents and insulin injections are the only way of treatment for the diabetes. Sometimes, synthetic hypoglycaemic agents cause serious side effects. Plant-based natural compounds may be best alternative of synthetic agents. Ethanolic extracts of *S. saponaria* have been detected as hypoglycaemic agent (Sahoo et al. 2010a, b). Alloxan-induced diabetic Wistar albino rats were used as experimental animal. Oral glucose tolerance and single-dose and multi-dose study was used for assessment of antidiabetic activity. Antidiabetic effect of extracts was also recorded from oral glucose tolerance test. Single-dose and multi-dose study of extracts also showed the significant antidiabetic activity and bring down the blood glucose level when compared to the diabetic control animals. The extract inhibited the elevation of glycosylated haemoglobin in vitro ($IC_{50} = 12.5\%$ g/ml). They also estimated the antioxidant enzymes such as SOD and catalase.

6.3 Pharmacological Properties of *S. saponaria*

S. saponaria is the second most explored species of the genus *Sapindus*. The different parts of *S. saponaria* possess various pharmacological activities such as antibacterial, antifungal, anti-leishmaniasis, anti-HIV and anti-snake venom activity. Most of the important activities are summarized in brief in Table 6.3. Here some of the activities are discussed in detail.

6.3.1 Anti-HIV Activity

Acquired immunodeficiency syndrome (AIDS) is the most deadly viral disease/ syndrome caused by the infection of human immunodeficiency virus (HIV). A number of plant's active molecules have been reported for their anti-HIV activity (Akanitapichat et al. 2002). Extracts were prepared from fresh *S. saponaria* fruits' peel using butanol, ethyl acetate, methanol and water, and used for the evaluation of anti-HIV activity (Rashed et al. 2013). Syncytia formation assay was used for the estimation of anti-HIV activity. Aqueous, butanol, ethyl acetate and methanolic extracts have been used for the evaluation of the anti-HIV-1 activity. The results showed that methanol extract has more potential as anti-HIV-1 agent with less EC_{50} and high therapeutic index. Fifty percent inhibition of syncytia formation has been observed from 91.852 μg/ml of methanolic extract.

6.3.2 Anti-snake Venom Activity

Among all the *Sapindus* species, only *S. saponaria* has been reported for their anti-snake venom activity. Leaves were cultured, and callus was used for the extracts' preparation into dicholoromethanol. These extracts were used for the fractionation through the TLC, and these fractions were used for the anti-snake venom activity (Silva et al. 2012). Male Swiss mice were used as experimental model. Mice were injected with the poison of different snakes (*B. jararacussu*, *B. moojeni*, *B. alternatus* and *C. d. terrificus*). For the assessment of anti-snake venom activity, haemorrhagic, myotoxic and phospholipase A_2 (PLA_2) activities were taken into account. Haemorrhagic activity is a predominant property of snake venom. PLA_2 is an important enzyme that breaks the phospholipids into lysophospholipids, and fatty acids finally ruptured the cell membrane. Most of the snake venom has the PLA_2-like activity and causes the muscle lysis. The myotoxic activity analysis is based on recognition of the enzyme creatine kinase released into the blood of the animal as the result of muscle lysis. Creatine kinase has been widely used as biomarker of muscle damage. Different extracts and fractions of *S. saponaria* moderately inhibited haemorrhagic, clotting, oedema-inducing, phospholipase and myotoxic activities of venom. Stigmasterol showed inhibitory properties on enzymatic and myotoxic activities of PLA_2.

Table 6.3 Pharmacological activities from *S. saponaria*

Biological activity	Plant parts	Extracts/ compounds	Microorganisms/details	References
Antioxidant				
Antimicrobial	Aerial	Hydromethanolic	*E. cloacae*, *E. coli*, *P. aeruginosa*, *S.* Typhimurium, *B. cereus*, *L. monocytogenes*, *M. flavus*, *S. aureus*	Rashed et al. (2013)
	Leaf	Methanolic, acetone, 1,4-dioxan	*Staphylococcus aureus*, *Proteus vulgaris, Pseudomonas aeruginosa and Micrococcus albus*	Prabhakar et al. (2012)
Antifungal	Dried pericarp	Aqueous, hydroalcoholic and butanolic	*C. parapsilosis*, *Candida albicans*, *C. glabrata*, *C. tropicalis* and *C. albicans* ATCC 90028	Tsuzuki et al. (2007)
	Aerial	Hydromethanolic	*A. fumigatus*, *A. versicolor*, *A. ochraceus*, *A. niger*, *T. viride*, *P. funiculosum*, *P. verrucosum*, *P. ochrochloron*	Rashed et al. (2013)
	Fruit's pericarp	Methanolic fraction of hydroethanolic extracts	*C. albicans*, *C. glabrata*	Damke et al. (2011)
Anticancer	Aerial	Hydromethanolic	MCF-7 (breast cancer adenocarcinoma), NCI-H460 (non-small cell lung cancer), HCT-15 (colon carcinoma), HeLa (cervical carcinoma) and HepG2 (hepatocellular carcinoma) cell lines	Rashed et al. (2013)
	Fruit's pericarp	Methanolic fraction of hydroethanolic extracts	HeLa (cervical carcinoma)	Damke et al. (2011)
Anti-*Trichomonas vaginalis*	Fruit's pericarp	Hydroethanolic, butanol extract, purified saponins	*Trichomonas vaginalis*	Damke et al. (2013)
Anti-ulcerogenic	Leaf	Hydroalcoholic extract	Reduce gastric secretion	Albiero et al. (2002)
Anti-snake venom	Callus of leaf	Dichloromethane-methanolic extracts fractions	Swiss albino mice was treated with venom of *B. jararacussu*, *B. moojeni*, *B. alternatus* and *C. d. terrificus*	Silva et al. (2012)
Larvicidal	Stem peel & seeds	Ethanolic extract	*Boophilus microplus and Rhipicephalus sanguineus*	Fernandes et al. (2005), Fernandes et al. (2007)
Spermicidal	Fruit's pericarp	Hydroethanolic, butanol extract, purified saponins	Human sperm	Damke et al. (2013)

6.3.3 Spermicidal and Anti-*Trichomonas* Activity

Sexually transmitted diseases are (STDs) most crucial public health problem and currently considered the key factor for HIV/AIDS transmission. Most of the family planning methods except condoms are effective against undesirable pregnancy, but not keep the protection against STDs. Protection from STDs and HIV infection can be minimized/stopped through the topical application of any substance that have the microbial and spermicidal activity. Damke et al. (2013) reported the spermicidal and anti-*Trichomonas* activity from the different extracts of *S. saponaria*. Dry pericarp of fruits was used for the preparation of water-ethanolic and butanolic extracts. These extracts have shown promising spermicidal and anti-*Trichomonas* activity. The 50% of spermatozoa was immobilized (EC_{50}) at the concentration of 0.5 (gram %) of water-ethanolic and saponins and 0.1 (gram %) of butanolic extracts. All these extracts/compounds were found effective against *T. vaginalis* with lower minimum inhibitory concentration (MIC) for clinical strain than ATCC strain.

6.3.4 Leishmanicidal and Trypanocidal Activity

Leishmaniasis and trypanosomiasis are major health problem issues. Both are parasitic diseases, having high infection rates. In Colombia fruit resin *S. saponaria* (Sapindaceae) is used to treat ulcers caused by leishmaniasis. Fruit's resin, chromatographical fractions of fruit resin and pure compounds isolated from extracts were tested against *Leishmania* species (*L. panamensis*, *L. braziliensis*, *L. amazonensis* and *L. donovani*) and *Trypanosoma cruzi*. They found promising leishmanicidal and trypanocidal activity from *S. saponaria* extracts (Correa et al. 2014). Crude extract would be a best alternative as leishmanicides, because of its natural origin, easy production, activity and minimal haemolytic effect.

Hydroethanolic extract, saponins and acrylic sesquiterpene oligoglycosides-rich fruit extracts of *S. saponaria* have been used to evaluate the anti-leishmaniasis activity (Moreira et al. 2017). Anti-leishmaniasis activity was evaluated using in vitro anti-proliferative assay, in vitro cytotoxicity and haemolytic assay. Mitochondrial dysfunction and phosphatidylserine exposure were investigated through flow cytometry. Of these used extracts/compounds, acrylic sesquiterpene oligoglycoside fraction showed the efficient leishmanicidal property.

6.3.5 Antimicrobial and Larvicidal Activity

Various concentrations of methanol, acetone and 1,4-dioxan extracts were tested for antimicrobial activity against *Staphylococcus aureus*, *Proteus vulgaris*, *Pseudomonas aeruginosa* and *Micrococcus albus*. All these extracts showed antimicrobial activity, but 1,4-dioxan extracts have more antimicrobial potential than other extracts (Prabhakar 2012).

Boophilus microplus is a common parasite of cattle. Stem peel of *S. saponaria* has been used for extraction in ethanol. Crude ethanolic extract has been used for larvicidal activity against larvae of *B. microplus*. Extract showed important larvicidal activity with LC_{50} (1258 ppm) and LC_{99} values (6360 ppm) (Fernandes et al. 2005).

Questions

1. Explain term saponins, its structure and the medicinal properties of saponins.
2. Explain the biosynthesis of saponins.
3. Describe the pharmacological activities of *S. mukorossi*.
4. Describe the pharmacological activities of *S. trifoliatus*.
5. Describe the pharmacological activities of *S. saponaria*.

Further Readings

Augustin, J. M., Kuzina, V., Anderson, S. B., & Bak, S. (2011). Molecular activities, biosynthesis an evolution of triterpenoid saponins. *Phytochemistry, 72*, 435–457.

Goyal, S., Kumar, D., Menaria, G., & Singla, S. (2014). Medicinal plants of the genus sapindus (sapindaceae) – A review of their botany, phytochemistry, biological activity and traditional uses. *Journal of Drug Delivery & Therapeutics, 4*, 7–20.

Upadhyay, A., & Singh, D. K. (2012). Pharmacological effects of *Sapindus mukorossi*. *Revista do Instituto de Medicina Tropical de São Paulo, 54*, 273–280.

Nanoparticles Synthesis and Nanotechnological Applications of *Sapindus* Species

7

7.1 Introduction

Nanotechnology is an emerging and continuously expanding area of research (Sergeev and Shabatina 2008). Nanotechnology promises with the synthesis and characterization of nanoparticles of desired shape and size by modulating of the materials up to the nanoscale. A wide range of applications of nanoparticles in the various fields promote the synthesis of nanoparticles (NPs). Nanoparticles are continuously used in the field of health care, biosensors, biotechnology, electronics, etc. (Mehata 2015; Ratnesh and Mehata 2015). Nanoparticle synthesis comprises the various chemical and physical methods. Green synthesis method is one of the most favoured methods for NP synthesis that is used in the health-care sector. Green synthesis method has been used for NP synthesis from plants (Makarov et al. 2014; Mittal et al. 2013), bacteria, fungus, yeasts (Narayanan and Sakthivel 2010) and DNA (Sohn et al. 2011). Green synthesis of NPs opens a new possibility that is relatively unexplored. Green synthesis method is more biocompatible and safe for medical use in comparison to the other physical and chemical methods (Verma and Mehta 2016).

Various chemical and physical methods have been developed and optimized for the production of NPs of the desired shape and size. However, these methods are costly and environmentally unfriendly. A wide range of applications supports the green synthesis method of NP production. Green synthesis method has been considered as one of the most encouraging methods for synthesis of nanoparticles due to their high biocompatibility, minimal toxicity and cost-effectiveness (Malik et al. 2014). The practice of plant extract for the NP synthesis (silver, gold, chitosan) deals with ample of benefits towards pharmaceutical and other health-care applications (Jain et al. 2009). A number of plants such as *Azadirachta indica*, *Aloe vera*, *Medicago sativa*, etc. and various parts of *Sapindus* species have been used for the nanoparticle synthesis through green synthesis method (Jha et al. 2009; Mittal et al.

R. Singh, B. Sharma, *Biotechnological Advances, Phytochemical Analysis and Ethnomedical Implications of Sapindus species*,
https://doi.org/10.1007/978-981-32-9189-8_7

2013; Makarov et al. 2014; Parashar et al. 2009). After the synthesis of NPs, characterization is required to determine the shape and size of NPs.

7.2 Nanoparticle Characterization

Nanoparticle characterization is a foremost stage to augment the nanoparticles synthesis. Suitable shape and size of nanoparticles improve the efficiency of activity. Several techniques are available for characterization of nanoparticles.

UV-Vis spectrophotometric method is a primary technique to characterize the NPs with scanning 300–700 nm. The UV-Vis absorption spectra showed the specific peaks for NPs. Silver NPs showed the specific peaks between 420 and 430 nm. X-ray diffraction (XRD) is another important technique for the characterization of NPs. The XRD peak pattern indicates the crystalline nature of NPs. Fourier transform infrared spectroscopy (FTIR) works on the principle that most molecules have the ability to absorb the infrared waves of electromagnetic spectrum. This amount of absorption directly correlates with the number of bonds present in the molecule. FTIR is a very sensitive technique to recognize the organic chemicals in a whole range of applications (Singh et al. 2014). Scanning electron microscopy (SEM) and transmission electron microscopy (TEM) are some of the best techniques to determine the NPs' shape and size.

7.3 Nanoparticle Synthesis from *Sapindus* Species

NPs have been prepared from the various parts of *Sapindus* species (Table 7.1).

7.3.1 *Sapindus rarak*

S. rarak rind aqueous extracts have been used for the preparation of zinc oxide (ZO) NPs. ZO NPs were prepared through hydrothermal methods, and these NPs have been characterized by XRD and SEM (Maryanti et al. 2014). Silver NPs have also been prepared using aqueous extract of *S. rarak* (Yudha et al. 2013). They also characterize the silver NPs using UV-Vis spectrophotometer and TEM.

7.3.2 *Sapindus mukorossi*

Different parts of *S. mukorossi* have been used for the synthesis of the NPs using various materials. Ramgopal and co-workers (2011) synthesized the silver NPs using the fruits of this plant. They evaluated the antimicrobial activity of these silver NPs against *Escherichia coli*, *Staphylococcus aureus*, *Pseudomonas aeruginosa* and *Serratia marcescens* and found the significant antimicrobial activity. Aqueous extract *S. mukorossi* fruit's pericarp was used as reducing agent as well as stabilizer

Table 7.1 Nanoparticle synthesis from *Sapindus* species using different materials

S. no.	Plant name	Plant part/ extracts	Metal/material	Activity	References
1	*Sapindus* spp.	Fruits	Silver	Antibacterial	Ramgopal et al. (2011)
2	*S. rarak*	Aqueous extract of rind	Silver	Not any	Yudha et al. (2013)
		Aqueous extract of rind	Zinc oxide	Not any	Maryanti et al. (2014)
3	*S. mukorossi*	Fruit pericarp extract	Gold	Reduction of p-nitroaniline	Reddy et al. (2013)
		Aqueous extract soapnut shells	Ammonium dihydrogen phosphate	Not any	Subha et al. (2015)
		Aqueous extract of fruit pericarp	Chitosan and silver	Antibacterial	Porchezhiyana and Noorjahan (2016)
		Fruit extract	Manganese oxide	Not any	Jassal et al. (2016)
		Seed extract	Palladium	Not any	Borah et al. (2017)
		Aqueous extract of fruit pericarp	Platinum	Not any	Kumar et al. (2017)

for synthesis of silver nanoparticles. This silver colloid showed antibacterial activity against *Bacillus subtilis*, *E. coli*, *S. aureus* and *P. aeruginosa*. This Ag colloid has also been evaluated for the degradation of 4-nitrophenol and Eosin blue (common environmental pollutants) (Dinda et al. 2017). Porchezhiyana and Noorjahan (2016) prepared the chitosan nanocomposite from silver nanoparticles of fruit pericarp of this plant. They standardized the formation of hydrogels and also found the significant antibacterial activity against *E. coli* and *S. aureus.*

Reddy et al. (2013) synthesized gold NPs (AuNPs) using fruit pericarp of *S. mukorossi* extract. They investigated the catalytic activity for reduction of p-nitroaniline. The fruit pericarp of *S. mukorossi* has also been used for the synthesis of platinum NPs (Kumar et al. 2017). Manganese NPs have been prepared using the fruits of *S. mukorossi* (Jassal et al. 2016). Palladium NPs from aqueous extract of *Sapindus mukorossi* seed have been synthesized by Borah et al. (2017).

Applications of Nanoparticles: Nanotechnology and nanoparticles have the wide range of application. These NPs showed high biological potential with applications in various fields such as agriculture, food science and technology, cosmetic, nanomedicine and health care sector. In agriculture, these NPs are used for controlled release of agrochemicals, detection and control of plant diseases, target-specific delivery of biomolecules, etc. NPs are extensively used in processing, storage and packaging processes in food science and technology. In bioengineering sector, NPs are an important part for the production of biocatalysts, photocatalysts, biosensors, etc. NPs are readily used in the production of sunscreen cream, anti-aging cream, hair growth products, bioactive compounds delivery, nanoemulsion, etc. Human health care sector is a prime area for the application of nanoparticles and nanotechnology. NPs are prominently used in the diagnostic as well as in therapeutics field. In therapeutics field, NPs are used as anti-oxidative, anti-microbial, antiproliferative, pro-apoptotic, anti-inflammatory activities, anti-parasitic, etc.

Questions

1. Explain the nanoparticles and its applications.
2. Describe the process of green synthesis method of nanoparticle production and advantages of this method.
3. List the different techniques that are used for the characterization of nanoparticles.

Further Readings

Dinda, G., Halder, D., Mitra, A., Pal, N., Vazquez-Vazquez, C., & Lapez-Quintela, M. A. (2017). Study of antibacterial and catalytic activity of silver colloid synthesized using fruit of *Sapindus mukorossi. New Journal of Chemistry, 41*, 10703–10711.

Porchezhiyana, V., & Noorjahan, S. E. (2016). Fabrication of a versatile chitosan nanocomposite hydrogel impregnated with biosynthesized silver nanoparticles using *Sapindus Mukorossi:* Characterization and applications. *RSC Advances, 6* (98), 95564–95573. https://doi.org/10.1039/C6RA19546F.

Bibliography

Abreu, A., Carulla, J. E., Lascano, C. E., Diaz, T. E., Kreuzer, M., & Hess, H. D. (2004). Effects of Sapindus saponaria fruits on ruminal fermentation and duodenal nitrogen flow of sheep fed a tropical grass diet with and without legume. *Journal of Animal Sciences, 82*, 1392–1400.

Adeyemi, T. O. (2011). *Molecular systematics and DNA barcoding of African Sapindaceae.* Unpublished PhD thesis, University of Lagos, Nigeria, pp 345.

Adeyemi, T. O., & Ogundipe, O. T. (2012). Biodiversity of Sapindaceae in West Africa: A checklist. *International Journal of Biodiversity and Conservation, 4*, 358–363.

Akanitapichat, P., Kurokawa, M., Tewtrakul, S., Pramyothin, P., Sripanidkulchai, B., Shiraki, K., & Hattori, M. (2002). Inhibitory activities of Thai medicinal plants against herpes simplex type 1, poliovirus type1, and measles virus. *Journal of Traditional Medicine, 19*, 174–180.

Akram, M., & Aftab, F. (2008). High frequency multiple shoot formation from nodal explants of teak (*Tectona grandis* L.) induced by Thidiazuran. *Propagation of Ornamental Plants, 8*, 72–75.

Alam, M. N., Bristi, N. J., & Rafiquzzaman, M. (2013). Review on in vivo and in vitro methods evaluation of antioxidant activity. *The Saudi Pharmaceutical Journal, 21*, 143–152.

Albiero, A. L. M., Bacchi, E. M., & Mourao, K. S. M. (2001). Caracterizacao anatomica das folhas, frutos esementes de *Sapindus saponaria* L. (Sapindaceae). *Acta Scientiarum: Biological Sciences, 23*, 549–560.

Albiero, A. L. M., Sertie, J. A. A., & Bacchi, E. M. (2002). Antiulcer activity of *Sapindus saponaria* L. in the rat. *Journal of Ethnopharmacology, 82*, 41–44.

Amin, M. N., & Jaiswal, V. S. (1987). Rapid clonal propagation of guava through *in vitro* shoot proliferation on nodal explants of mature trees. *Plant Cell, Tissue and Organ Culture, 9*, 235–243.

Ammirato, P. V. (1983). Embryogenesis. In D. A. Evans, W. R. Sharp, P. V. Ammirato, & Y. Yamada (Eds.), *Handbook of plant cell culture* (Vol. 1, pp. 82–123). New York: Mac Millan Publishing Co.

Aneja, K. R., Joshi, R., & Sharma, C. (2010). *In vitro* antimicrobial activity of *Sapindus mukorossi* and *Emblica officinalis* against dental caries pathogens. *Ethanobotanical Leaflets, 14*, 402–412.

Andrade, G., Shah, R., Johansson, S., & Egertsdotter, U. (2011). Somatic embryogenesis as a tool for forest tree improvement: A case study in Eucalyptus globules. *BMC Proceedings, 5*. https://doi.org/10.1186/1753-6561-5-S7-P128.

Angiosperm Phylogeny Group II. (2003). An update of the Angiosperm Phylogeny Group classification for the orders and families of flowering plants: APG II. *Botanical Journal of the Linnean Society, 141*, 399–436.

Anonymous. (1972). *The wealth of India. Industrial products Vol. IX* (pp. 222–229). New Delhi: Publication and Information Directorate, CSIR.

Anonymous. (1981). *Biotechnology: A Dutch perspectives*. A report by the Netherlands Study Centre for Techology Trends. JHF Van Apeldoorn (Ed.).

R. Singh, B. Sharma, *Biotechnological Advances, Phytochemical Analysis and Ethnomedical Implications of Sapindus species*,
https://doi.org/10.1007/978-981-32-9189-8

Anonymous. (1992). The useful plants of India, A dictionary of Indian Raw Material and Publications and Information Directorate CSIR, New Delhi.

Ara, H., Jaiswal, U., & Jaiswal, V. S. (2000). Somatic embryogenesis and regeneration in amrapali and chausa cultivars of mango (Mangifera indica L.). *Current Sciences, 78*, 164–169.

Arnold, S. V., & Hakman, I. (1988). Regulation of somatic embryo development in Picea abies by abscisic acid (ABA). *Journal of Plant Physiology, 132*, 164–169.

Arora, B., Bhadauria, P., Tripathi, D., & Sharma, A. (2012). *Sapindus emarginatus*: Phytochemistry & various biological activities. *Indo Global Journal of Pharmaceutical Sciences, 2*, 250–257.

Arul, B., Kothai, R., & Jacob, P. (2004). Anti-inflammatory activity of *Sapindus trifoliatus* Linn. *Journal of Herbal Pharmacotherapy, 4*, 43–50.

Arulmozhi, D. K., Veeranjaneyulu, A., Bodhankar, S. L., & Arora, S. K. (2004). Investigations into the antinociceptive activity of *Sapindus trifoliatus* in various pain models. *The Journal of Pharmacy and Pharmacology, 56*, 655–661.

Arulmozhi, D. K., Veeranjaneyulu, A., Bodhankar, S. L., & Arora, S. K. (2005a). Pharmacological investigations of *Sapindus trifoliatus* in various in-vitro and in-vivo models of inflammation. *Indian Journal of Pharmacology, 37*, 96–102.

Arulmozhi, D. K., Veeranjaneyulu, A., Bodhankar, S. L., & Arora, S. K. (2005b). Effect of *Sapindus trifoliatus* on hyperalgesic in-vivo migraine models. *Brazilian Journal of Medical and Biological Research, 38*, 469–475.

Arulmozhi, D. K., Veeranjaneyulu, A., Bodhankar, S. L., & Arora, S. K. (2005c). Investigations of *Sapindus trifoliatus* in dopaminergic and serotonergic systems, purative anti migraine mechanisms. *Indian Journal of Pharmacology, 37*, 120–125.

Aslam, J., Mujib, A., & Sharma, M. P. (2014). Somatic embryos in Catharanthus roseus: A scanning electron microscopic study. *Notulae Scientia Biologicae, 6*, 167–172.

Asthana, P. (2010). *Morphogenetic studies in Sapindus trifoliatus L.* PhD Thesis, Banaras Hindu University, Varanasi, UP, India.

Asthana, P., Jaiswal, V. S., & Jaiswal, U. (2011). Micropropagation of Sapindus trifoliatus L. and assessment of genetic fidelity of micropropagated plants using RAPD analysis. *Acta Physiologae Plantarum, 33*, 1821–1829.

Asthana, P., Rai, M. K., & Jaiswal, U. (2017). Somatic embryogenesis from sepal explants in *Sapindus trifoliatus*, a plant valuable in herbal soap industry. *Industrial Crops and Products, 100*, 228–235.

Augustin, J. M., Kuzina, V., Anderson, S. B., & Bak, S. (2011). Molecular activities, biosynthesis an evolution of triterpenoid saponins. *Phytochemistry, 72*, 435–457.

Azad, M. A. K., Yokota, S., Begum, F., & Yoshizawa, Y. (2009). Plant regeneration through somatic embryogenesis of a medicinal plant, Phellodendron amurense Rupr. *In Vitro Cellular & Developmental Biology-Plant, 45*, 441–449.

Azhar, I., Usmanghani, K., Perveen, S., Ali, M. S., & Ahmad, V. U. (1993). Two triterpenoidal saponins from *S. mukorossi* Gaertn. *Pakistan Journal of Pharmaceutical Sciences, 6*, 71–77.

Azhar, I., Usmanghani, K., Perveen, S., Ali, M. S., & Ahmad, V. U. (1994). Chemical constituents of *S. mukorossi* Gaertn. *Pakistan Journal of Pharmaceutical Sciences, 7*, 33–41.

Baker, J. T., Borris, R. P., Carte, B., Cordell, G. A., Soejarto, D. D., Cragg, G. M., Gupta, M. P., Iwu, M. M., Madulid, D. R., & Tyler, V. E. (1995). Natural product drug discovery and development: New perspective on international collaboration. *Journal of Natural Product, 58*, 1325–1357.

Benelli, C., Fabbri, A., Grassi, S., Lambardi, M., & Rugini, E. (2001). Histology of somatic embryogenesis in mature tissue of olive (*Olea europaea* L.). *The Journal of Horticultural Science and Biotechnology, 76*, 112–119.

Berthoulay, M., & Michaux-Ferriere, N. M. (1996). High frequency somatic embryogenesis in Coffea canephora. Induction conditions and histological evolution. *Plant Cell Tissue and Organ Culture, 44*, 169–176.

Bhargava, D., Shivapuri, J. N., Kar, S., Pandit, B. R., Sidhiqie, A., Upadhayay, A., Thakur, S., & Mondal, K. C. (2012). Evaluation of antigonorrhoeal activity of saponins extract of *Sapindus mukorossi* Gaertn. *Research Journal of Pharmaceutical, Biological and Chemical Sciences, 3*, 459–470.

Bhansali, R., Driver, J. A., & Durzan, D. J. (1991). Adventitious embryogenesis and plant regeneration from rescued embryos of peach, *Prunus persica* (L.). *The Indian Journal of Experimental Biology, 29*, 334–337.

Bhojwani, S. S., & Dantu, P. K. (2013). *Plant tissue culture- an introductory, text*. Springer. https://doi.org/10.1007/978-81-322-1026-9.

Bhojwani, S. S., & Razdan, M. K. (1996). *Plant tissue culture: Theory and practice*, A revised edition. Amsterdam: Elsevier.

Bodhankar, S. L., Garg, S. K., & Mathur, V. S. (1974). Antifertility screening of plants. Part IX. Effect of five indigenous plant on early pregnanacy in female albino rats. *Indian Journal of Medical Research, 62*, 831–837.

Bonga, J. M. (1981). Vegetative propagation of mature trees by tissue culture. In A. N. Rao (Ed.), *Proceedings of the symposium on tissue culture of economoically important plants* (pp. 124–137). Singapore: COSTED/ANBS.

Borah, R. K., Mahanta, A., Dutta, A., Bora, U., & Thakur, A. J. (2017). A green synthesis of palladium nanoparticles by *Sapindus mukorossi* seed extract and use in efficient room temperature Suzuki–Miyaura cross-coupling reaction. *Applied Organometallic Chemistry, 31*(11), 3784. https://doi.org/10.1002/aoc.3784.

Buendía-González, L., Estrada-Zúñiga, M. E., Orozco-Villafuerte, J., Cruz-Sosa, F., & Vernon-Carter, E. J. (2012). Somatic embryogenesis of the heavy metal accumulator Prosopis laevigata. *Plant Cell Tissue and Organ Culture, 108*, 287–296.

Buitink, J., & Leprince, O. (2008). Intracellular glasses and seed survival in the dry state. *Comptes Rendus Biologies, 331*, 788–795.

Bunn, E. (2005). Development of *in vitro* methods for *ex situ* conservation of *Eucalyptus impensa*, an endangered mallee from Southwest Western Australia. *Plant Cell, Tissue and Organ Culture, 83*, 97–102.

Capuana, M., & Debergh, P. C. (1997). Improvement of the maturation and germination of horse chestnut somatic embryos. *Plant Cell Tissue and Organ Culture, 48*, 23–29.

Ceasar, S. A., Maxwell, S. L., Prasad, K. B., Karthigan, M., & Ignacimuthu, S. (2010). Highly efficient shoot regeneration of *Bacopa monnieri* L. using a two-stage culture procedure and assessment of genetic integrity of micropropagated plants by RAPD. *Acta Physiologiae Plantarum, 32*, 443–452.

Chakraborty, A., Amudha, P., Geetha, M., Surjit, & Singh, N. (2010). Evaluation of anxiolytic activity of methanolic extract of *Sapindus mukorossi* Gaertn in mice. *International Journal of Pharma and Bio Sciences, 1*, 1–8.

Chandrasekhar, T., Hussain, T. M., & Jayanand, B. (2005). *In vitro* Micropropagation of *Boswellia ovalifoliolata*. Verlag der Zeitschrift für Naturforschung, Tübingen. http://www.znaturforsch.com

Chappell, J. (2002). The genetics and molecular genetics of terpene and sterol origami. *Current Opinion in Plant Biology, 5*, 151–157.

Chattopadhyay, D., Arunachalam, G., Mandal, A. B., Bhadra, R., & Mandal, S. C. (2003). CNS activity of *Mallotus peltatus* Muell Arg. leaf extracts. An ethnomedicine Onge. *Journal of Ethnopharmacology, 85*, 99–105.

Chandrika, M., & Rai, R. V. (2009). Genetic fidelity in micropropagated plantlets of Ochreinauclea missionis an endemic, threatened and medicinal tree using ISSR markers. *African Journal of Biotechnology, 8*, 2933–2938.

Chen, Y. H., Chiang, T. H., & Chen, J. H. (2013). Properties of soapnut (*Sapindus mukorossi*) oil biodiesel and its blends with diesel. *Biomass and Bioenergy, 52*, 15–21.

Chen, C. Y., Kuo, P. L., Chen, Y. H., Huang, J. C., Ho, M. L., & Lin, R. J. (2010). Tyrosinase inhibition, free radical scavenging, anti-microorganism and anti-cancer proliferation activities of *Sapindus mukorossi extracts. J Taiwan Inst Chem Engrs, 41,* 129–135.

Cheng, X. Y., Wei, T., Guo, B., Ni, W., & Liu, C. Z. (2005). *Cistanche deserticola* cell suspension cultures: Phenylethanoid glycosides biosynthesis and antioxidant activity. *Process Biochemistry, 40*, 3119–3124.

Chhetri, A. B., Tango, M. S., Budge, S. M., Watts, K. C., & Islam, M. R. (2008). Non-edible plant oils as new sources for biodiesel production. *International Journal of Molecular Sciences, 9*, 169–180.

Chirva, V. Y., Kintya, P. K., Sosnovskii, V. A., Krivenchuk, P. E., & Zykova, N. Y. (1970). A tri terpene glycosides of *S. mukorossi*-D part 2 structure of sapindoside A and sapindoside B. *Khim Prir Soedin (Tashk), 6*, 218–221.

Chirva, V. Y., Kintya, P. K., Sosnovskii, V. A., & Zolotarev, B. M. (1973a). Tri terpene glycosides of *S.mukorossi* part 3 structure of sapindoside C. *Chem of Natural Compounds (English Transltion of Khim Prir Soedin), 6*, 380–381.

Chirva, V. Y., Kintya, P. K., Sosnovskii, V. A., & Zolotarev, B. M. (1973b). Triterpene glycosides of *S.mukorossi* part 5 structure of sapindoside D. *Khim Prir Soedin (Tashk), 6*, 316–319.

Chirva, V. Y., Kintya, P. K., Sosnovskii, V. A., & Zolotarev, B. M. (1973c). Triterpene glycosides of *S.mukorossi* part 5 structure of sapindoside E. *Chemistry of Natural Compounds (English Transltion of Khim Prir Soedin), 6*, 440–442.

Chuang, S. J., Chen, C. L., Chen, J. J., Chou, W. Y., & Sung, J. M. (2009). Detection of somaclonal variation in micro-propagated *Echinacea purpurea* using AFLP marker. *Scientia Horticulturae, 120*, 121–126.

Chung, S. J., Cen, L. C., Chen, J. J., Chou, W. Y., & Sung, J. M. (2009). Detection of somaclonal variations in micro-propagated Echinacea purpurea using AFLP marker. *Scientia Horticulturae, 120*, 121–126.

Cordell, G. A. (1995). Changing strategies in natural products chemistry. *Phytochemistry, 40*, 1585–1612.

Coste, A., Oltean, B., Halmagyi, A., & Deliu, C. (2011). Direct somatic embryogenesis and plant regeneration in Peucedanum oreoselinum (L.) Moench. *Romanian Biotechnological Letters, 16*, 6450–6459.

Correa, E., Quiñones, W., Robledo, S., Carrillo, L., Archbold, R., Torres, F. N., Escobar, G., Herrera, N., & Echeverri, F. (2014). Leishmanicidal and trypanocidal activity of *Sapindus saponaria. Boletin Latinoamericano y del Caribe de Plantas Medicinales y Aromaticas, 13* (4), 311–323.

Corredoira, E., San-Jose, M. C., & Vieitez, A. M. (2012). Induction of somatic embryogenesis from different explants of shoot cultures derived from young Quercus alba trees. *Trees Structure and Function, 26*, 881–891.

Damke, E., Tsuzuki, J. K., Cortez, D. A. G., Ferreira, I. C. P., Bertoni, T. A., Batista, M. R., Donati, L., Svidzinski, T. I. E., & Consolaro, M. E. L. (2011). In vivo activity of Sapindus saponaria against azole-susceptible and -resistant human vaginal Candida species. *BMC Complementary and Alternative Medicine, 11*, 35.

Damke, E., Tsuzuki, J. K., Chassot, F., Cortez, D. A. G., Ferreira, I. C. P., Mesquita, C. S. S., Da-Silva, V. R. S., Svidzinski, T. I. E., & Consolaro, M. E. L. (2013). Spermicidal and anti-Trichomonas vaginalis activity of Brazilian *Sapindus saponaria. BMC Complementary and Alternative Medicine, 13*, 196. https://doi.org/10.1186/1472-6882-13-196.

Debergh, P. C., & Read, P. E. (1991). Micropropagation. In P. C. Debergh & R. H. Zimmerman (Eds.), *Micropropagation: Technology and application* (pp. 1–13). Dordrecht: Kluwer Academic Publisher.

Desai, H. V., Prashant, N. B., & Mehta, A. R. (1986). Plant regeneration of *Sapindus trifoliatus* L. (Soapnut) through somatic embryogenesis. *Plant Cell Reports, 3*, 190–191.

Desai, H. V., Prashant, N. B., & Mehta, A. R. (1986). Plant regeneration of Sapindus trifoliatus L. (Soapnut) through somatic embryogenesis. *Indian Journal of Experimental Biology, 3*, 190–191.

Dev, I., & Guha, S. R. D. (1979). Glyceride composition of *Sapindus mukorossi* (soapnut) oil. *Indian Journal of Forestry, 2*, 261–263.

Dhar, U., & Upreti, J. (1999). *In vitro* regeneration of a mature tree leguminous liana (*Bauhinia vahlii* wight and Arnott). *Plant Cell Reports, 18*, 664–669.

Dhar, J. P., Bajpai, V. K., Setty, B. S., & Kamboj, V. P. (1989). Morphological changes in human spermatozoa as examined under scanning electron microscope after in vitro exposure to saponins isolated from *Sapindus mukorossi*. *Contraception, 39*, 563–568.

Dinda, G., Halder, D., Mitra, A., Pal, N., Vazquez-Vazquez, C., & Lopez-Quintela, M. A. (2017). Study of antibacterial and catalytic activity of silver colloid synthesized using fruit of *Sapindus mukorossi*. *New Journal of Chemistry, 41*, 10703–10711.

Dixit, V. P., & Gupta, R. S. (1982). Antispermatogenic and antiandrogenic activity of *Sapindus trifoliatus* fruit extract in intact and castrated male gerbils. *Planta Medica, 46*, 242–246.

Dobelis, I. N. (1993). *Magic and medicine of plants* (pp. 8–48). Pleasant/New York: The Readers Digest Association Inc.

Dobhal, U., Bhandari, S., Bisht, S., & Bisht, N. S. (2012). Plant regeneration through somatic embryogenesis in *Sapindus mukorossi*. *Plant Archives, 12*, 219–221.

Dodeman, V. L., Ducreux, G., & Kreis, M. (1997). Zygotic embryogenesis *versus* somatic embryogenesis. *Journal of Experimental Botany, 48*, 1493–1509.

Doyle, J. J., & Doyle, J. L. (1990). Isolation of plant DNA fresh tissue. *Focus, 12*, 13–15.

Dudits, D., Gyorgyey, J., Bogre, L., & Bako, L. (1995). Molecular biology of somatic embryogenesis. In T. A. Thorpe (Ed.), *In vitro embryogenesis in plants* (pp. 267–308). Dordrecht: Kluwer Academic Publisher.

Duncan, D. B. (1955). Multiple range and multiple F test. *Biometrics, 11*, 1–42.

Emons, A. M. C. (1994). Somatic embryogenesis, cell biological aspects. *Acta Botanica Neerlandica, 43*, 1–14.

Eklund, P. C., Langvik, O. K., Warna, J. P., Salmi, T. O., Willfor, S. M., & Sjoholm, R. E. (2005). Chemical studies on antioxidant mechanisms and free radical scavenging properties of lignans. *Organic and Biomolecular Chemistry, 21*, 3336–3347.

Erdogrul, O. T. (2002). Antibacterial activities of some plant extract used in folk medicine. *Pharmaceutical Biology, 40*, 269–273.

Fajriaty, I., Ketut, K., & Fidrianny, A. (2014). Acute and sub-chronic (28 days) repeated oral toxicity test of ethanol extract of lerak (Sapindus rarak. dc) fruits in Wistar rats. *International Journal of Pharmacy and Pharmaceutical Sciences, 6*, 487–492.

Fan, C. Q., & Yue, J. M. (2003). Biologically active phenols from *Saussurea medusa*. *Bioorganic & Medicinal Chemistry, 11*, 703–708.

Fauconneau, B., Waffo-teguop, P., Huguet, F., Barrier, L., Decendit, A., & Mjzrillon, J. M. (1997). Comparative study of radical scavenger and antioxidant properties of phenolic compounds from *Vitis vinifera* cell cultures using in vitro test. *Life Sciences, 61*, 2103–2110.

Feher, A., Pasternak, T. P., & Dudits, D. (2003). Transition of somatic plant cells to an embryogenic state. *Plant Cell Tissue and Organ Culture, 74*, 201–228.

Fernandes, F. F., Freitas, E. P. S., Costa, A. C., & Silva, I. G. (2005). Larvicidal potential of Sapindus saponaria to control of the cattle tick Boophilus microplus. *Pesquisa Agropecuaria Brasileira, 40*, 1243–1245.

Fernandes, F. F., Leles, R. N., Silva, I. G., & Freitas, E. P. S. (2007). Larvicidal potencial of Sapindus saponaria (Sapindaceae) against Rhipicephalus sanguineus (Latreille, 1806) (Acari: Ixodidae). *Arquivo Brasileiro de Medicina Veterinaria e Zootecnia, 59*, 145–149.

Ferrucci, M. S. (2000). Cytotaxonomy of Sapindaceae with special reference to the tribe Paullinieae. *Genetics and Molecular Biology, 23*, 941–946.

Fitch, M., & Manshardt, R. M. (1990). Somatic embryogenesis and plant regeneration from immature zygotic embryos of papaya (Carica papaya L.). *Plant Cell Reports, 9*, 320–324.

Francis, G., Kerem, Z., Makkar, H., & Becker, K. (2002). The biological action of saponins in animal systems: A review. *The British Journal of Nutrition, 88*, 587–605.

George, B., & Shanmugam, S. (2014). Phytochemical screening and antimicrobial activity of fruits extract of *Sapindus mukorossi*. *International Journal of Current Microbiology and Applied Sciences, 3*, 604–611.

George, E. F., Hall, M. A., & Klerk, J. D. (2008). Plant propagation by tissue culture. *The Background Springer, 1*, 65–75.

Ghani, A. (2003). *A medicinal plants of Bangladesh with chemical constituents and uses* (p. 398). Dhaka: The Asiatic Society of Bangladesh.

Gill, M. S., Singh, Z., Dhillon, B. S., & Ghoshal, S. S. (1995). Somatic embryogenesis and plantlet regeneration in mandarin (Citrus reticulate Blanco). *Scientia Horticulturae, 63*, 167–174.

Giri, C. C., Shyamkumar, B., & Anjaneyulu, C. (2004). Progress in tissue culture, genetic transformation and applications of biotechnology to trees: An overview. *Trees – Structure and Function, 18*, 115–135.

Giri, L., Dhyani, P., Rawata, S., Bhatta, I. D., Nandia, S. K., Rawala, R. S., & Pande, V. (2012). In vitro production of phenolic compounds and antioxidant activity in callus suspension cultures of *Habenaria edgeworthii:* A rare Himalayan medicinal orchid. *Industrial Crops and Products, 39*, 1–6.

Gordon, M. H. (1990). The mechanism of antioxidant action in vitro, in food antioxidants. In B. J. F. Hudson (Ed.), *Applied sciences* (pp. 1–18). London: Elesvier.

Goyal, S., Kumar, D., Menaria, G., & Singla, S. (2014). Medicinal plants of the genus sapindus (sapindaceae) – A review of their botany, phytochemistry, biological activity and traditional uses. *Journal of Drug Delivery and Therapeutics, 4*, 7–20.

Grover, R. K., Roy, A. D., Roy, R., Joshi, S. K., Srivastava, V., Sudershan, K., & Arora, S. K. (2005). Complete 1H and 13C NMR assignments of six saponins from Sapindus trifoliatus. *Magnetic Resonance Chemistry, 43*, 1072–1076.

Gulcin, I., Berashvili, D., & Gepdiremen, A. (2005). Antiradical and antioxidant activity of total anthocyanins from *Perilla pankenensis* Decne. *The Journal of Ethnopharmacology, 101*, 287–293.

Gupta, D. R., & Ahmed, B. (1990). Emarginatosides B and C: Two new saponins from *Sapindus emarginatus* fruits. *Indian Journal of Chemistry, 29B*, 268–270.

Gupta, P. K., & Durzan, D. J. (1987). Biotechnology of somatic polyembryogenesis and plantlet regeneration in loblolly pine. *Biotechnology, 5*, 147–151.

Gupta, A. K., Harish, R. M. K., Phulwaria, M., Agarwal, T., & Shekhawat, N. S. (2014). In vitro propagation, encapsulation, and genetic fidelity analysis of Terminalia arjuna: A cardioprotective medicinal tree. *Applied Biochemistry and Biotechnology, 173*, 1481–1494.

Hamburger, M., Scalanin, I., Hostettmann, K., Dyatmiko, W., & Sutarjadi. (2007). Acetylated saponins with molluscicidal activity from Sapindus rarak: Unambiguous structure determination by proton nuclear magnetic resonance and quantitative analysis. *Phytochemical Analysis, 3*, 231–237.

Hammer, K. A., Carson, C. F., & Riley, T. V. (1999). Antimicrobial activity of essential oils and other plant extracts. *Journal of Applied Microbiology, 86*, 985–990.

Hammerschlag, F. A. (1986). Temperate fruits and nuts. In R. H. Zimmerman, R. J. Griesbach, R. J. Griesbach, F. A. Hammerschlag, & R. H. Lawson (Eds.), *Tissue culture as a plant production system for horticultural crops* (pp. 221–236). Dordrecht: Martinus Nijhoff Publishers.

Havesteen, B. (1983). Flavonoids, a class of natural products of high pharmacological potency. *Biochemical Pharmacology, 32*, 1141–1148.

Havsteen, B. (1983). Flavonoids, a class of natural products of high pharmacological potency. *Biochemical Pharmacology, 32*, 1141–1148.

Hegazi, G. A. E. (2011). In vitro studies on *Delonix elata* L. – An endangered medicinal plant. *World Applied Sciences Journal, 14*, 679–686.

Hempel, J., Pforte, H., Raab, B., Engst, W., Bohm, H., & Jacobasch, G. (1999). Flavonols and flavones of parsley cell suspension culture change the antioxidative capacity of plasma in rats. *Nahrung, 43*, 201–204.

http://onlinelibrary.wiley.com/doi/10.1002/pca.2800030507/abstract. Diakses 2 April.

Hu, Q., Chen, Y. Y., Jiao, Q. Y., Khan, A., Li, F., Han, D. F., Cao, G. D., & Lou, H. X. (2018). Triterpenoid saponins from the pulp of Sapindus mukorossi and their antifungal activities. *Phytochemistry, 147*, 1–8.

Huang, C., Liao, S. C., Chang, F. R., Kuo, Y. H., & Wu, Y. C. (2003). Molluscicidal Saponins from *Sapindus mukorossi. Journal of Agricultural and Food Chemistry, 51*, 4916–4919.

Huang, D., Ou, B., & Prior, R. L. (2005). The chemistry behind antioxidant capacity assays. *Journal of Agricultural and Food Chemistry, 53*, 1841–1856.

Huang, H. C., Tsai, W. J., Morris-Natschke, S. L., Tokuda, H., Lee, K. H., & Wu, Y. C. (2006). Sapinmusaponins F-J, bioactive tirucallane-type saponins from the galls of *Sapindus mukorossi. Journal of Natural Products, 69*, 763–767.

Huang, H. C., Tsai, W. J., Liaw, C. C., Wu, S. H., Wu, Y. C., & Kuo, Y. H. (2007). Anti-platelet aggregation triterpene saponins from the galls of *Sapindus mukorossi. Chemical & Pharmaceutical Bulletin (Tokyo), 55*, 1412–1415.

Ibrahim, M., Khan, A. A., Tiwari, S. K., Habeeb, M. A., Khaja, M. N., & Habibullah, C. M. (2006). Anti-microbial activity of *Sapindus mukorossi* and *Rheum emodi* extracts against *Helicobacter pylori: In vitro* and *in vivo* studies. *World Journal of Gastroenterology, 12*, 7136–7142.

Ibrahim, M., Khaja, M. N., Aara, A., Khan, A. A., Habeeb, M. A., Devi, Y. P., et al. (2008). Hepatoprotective activity of *Sapindus mukorossi* and *Rheum emodi* extracts: *In vitro* and *in vivo* studies. *World Journal of Gastroenterology, 14*, 2566–2571.

Jack, O.-O. (2008). Phytochemical analysis and antimicrobial activity of the leaves of fleabane (*Conyza sumatrensis*). *Journal of Applied Sciences and Environmental Management, 12*, 63–65.

Jain, D., Daima, H. K., Kachhwaha, S., & Kothari, S. L. (2009). Synthesis of plant mediated silver nanoparticles using papaya fruit extract and evaluation of their antimicrobial activities. *Digest Journal of Nanomaterials and Biostructures, 4*, 557–563.

Janero, D. R. (1990). Malondialdehyde and thiobarbituric acidreactivity as diagnostic indices of lipid peroxidation and peroxidative tissue injury. *Free Radical Biology & Medicine, 9*, 515–540.

Jaiswal, V. S. (2003). Organogenic and embryogenic regeneration and their role in in vitro conservation of tropical fruit species. In R. Chaudhury, R. Pandey, S. K. Malik, & B. Mal (Eds.), *In vitro conservation and cryopreservation of tropical fruit species* (pp. 101–113). IPGRI Office for South Asia and NBPGR: New Delhi.

Jassal, V., Shanker, U., Gahlot, S., Kaith, B. S., Kamaluddin, Iqubal, M. A., & Samue, P. (2016). Sapindus mukorossi mediated green synthesis of some manganese oxide nanoparticles interaction with aromatic amines. *Applied Physics A: Materials Science & Processing, 122*, 271.

Jayasree, T., Basha, G., Chandrasekhar, N., Iqbal, S. U., Dixit, R., & Manohar, V. S. (2011). Evaluation into antiepileptic activity of aqueous extract of *Sapindus trifoliatus* (pericarp) in Swiss albino rats. *Asian Journal of Research in Biological and Pharmaceutical Sciences, 1*, 459–463.

Jayasree, T., Naveen, A., Chandrasekhar, N., Sunil, M., Kishan, P. V., & Jagan, R. N. (2012). Evaluation of muscle relaxant activity of aqueous extract of *Sapindus trifoliatus* (pericarp) in swiss albino mice. *Journal of Chemical and Pharmaceutical Research, 4*, 1960–1964.

Jeyabalan, S., & Palayan, M. (2009). Antihyperlipidemic activity of *Sapindus emarginatus* in Triton WR-1339 induced albino rats. *Research Journal of Pharmacy and Technology, 2*, 319–323.

Jha, A. K., Prasad, K., Prasad, K., & Kulkarni, A. R. (2009). Plant system: Nature's nanofactory. *Colloids and Surfaces B: Biointerfaces, 73*, 219–223.

Jimenez, V. M., Guevara, E., Herrera, J., & Bangerth, F. (2001). Endogenous hormone levels in habituated nucellar *Citrus* callus during the initial stages of regeneration. *Plant Cell Reports, 20*, 92–100.

Kaladhar, D. S. V. G. K., Geetha, S. M., Varahalarao, V., & Nagendra, S. Y. (2013). Anti-dandruff activity of ethanolic extract of *Sapindus mukorossi* seed coat and *Ficus racemosa* fruit peel and in silico protein interaction studies. *International Journal of Advanced Pharmaceutical Sciences, 2*, 149–155.

Kalinowska, M., Zimowski, J., Paczkowski, C., & Wojciechowski, Z. A. (2005). The formation of sugar chains in triterpenoid saponins and glycoalkaloids. *Phytochemistry Reviews, 4*, 237–257.

Kanchanapoom, T., Kasai, R., & Yamasaki, K. (2001). Acetylated triterpene saponins from the Thai medicinal plant *Sapindus emarginatus. Chemical & Pharmaceutical Bulletin, 49*, 1195–1197.

Kasai, R., Fujino, H., Kuzuki, T., Wong, W. H., Goto, C., & Yata, N. (1986). Acyclic sesquiterpene oligoglycosides from pericarps of *Sapindus mukorossi*. *Phytochemistry, 25*, 871–876.

Kasai, R., Nishi, M., Mizutani, K., Miyahara, I., Moriya, T., Miyahara, K., & Tanaka, O. (1988). Trifolioside II an acrylic sesquiterpene oligoglycoside from pericarp of S. trifoliatus. *Phytochemistry, 27*, 2309–2311.

Keser, S., Celik, S., Turkoglu, S., Yilmaz, O., & Turkoglu, I. (2012). Hydrogen peroxide radical scavenging and total antioxidant activity of Hawthorn. *Chemistry Journal, 2*, 9–12.

Khatamian, H., & Abuelgasim, Z. (1986). Auxins aid soapberry cuttings. *American Nurseryman, 164*, 65–72.

Khlifi, S., & Tremblay, F. M. (1995). Maturation of black spruce somatic embryos. Part I. effect of L-glutamine on the number and germinability of somatic embryos. *Plant Cell Tissue and Organ Culture, 41*, 23–32.

Kim, S., Yang, M., Lee, O., & Kang, S. (2011). The antioxidant activity and the bioactive compound content of *Stevia rebaudiana* water extract. *Food Science and Technology, 44*, 1328–1332.

Kim, H. T., Yang, B. H., Park, Y. G., & Liu, J. R. (2012). Somatic embryogenesis in leaf tissue culture of soapberry (Sapindus mukorossi Gaertn). *Plant Biotechnology, 29*, 311–314.

Kimata, H., Nakashima, T., Kokubun, S., Nakayama, K., Mitima, Y., Kitahara, T., & Tanaka, O. (1983). Saponins of pericarps of *Sapindus mukorossi* Garetn. and solubilisation of monodesmosides by bisdesmosides. *Chemical & Pharmaceutical Bulletin, 31*, 1998–2005.

Kirby, J., & Keasling, J. D. (2009). Biosynthesis of plant isoprenoids: Perspectives for microbial engineering. *Annual Review of Plant Biology, 60*, 335–355.

Kirtikar, K. R., & Basu, B. D. (1991). *Indian medicinal plants*. Allahabad: BLM Publication.

Kirtikar, K. R., & Basu, B. D. (1999). *Indian medicinal plants 1* (pp. 632–635). New Delhi: International Book Distributors.

Kishore, D. V., Pinto, J., Mini, K. V., Ronald, F., & Satyanarayana, D. (2010). Anti-ulcer activity of leaf extract of *Sapindus trifoliatus* Linn. *International Journal of Asia Pacific Studies, 1*, 104–107.

Koleva, I. I., Van Beek, T. A., Linseen, J. P. H., De Groot, & Evstatieva, L. N. (2002). Screening of plant extracts for antioxidant activity: A comparative study on three testing methods. *Phytochemical Analysis, 10*, 178–182.

Komamine, A., Murata, N., & Nomura, K. (2005). Mechanisms of somatic embryogenesis in carrot suspension cultures–morphology, physiology, biochemistry and molecular biology. *In Vitro Cellular and Developmental Biology-Plant, 41*, 6–10.

Kommalapati, R. R., Valsaraj, K. T., Constant, W., & Roy, D. (1998). Soil flushing using colloidal gas aphron suspensions generated from a plant-based surfactant. *The Journal of Hazardous Materials, 60*, 73–87.

Kovacheva, E., Georgiev, M., Pashova, S., Angelova, M., & Ilieva, M. (2006). Radical quenching by rosmarinic acid from *Lavandula vera* MM cell culture. *Zeitschrift für Naturforschung, 61C*, 517–520.

Krishnan, P. N., Decruse, S. W., & Radha, R. K. (2011). Conservation of medicinal plants of Western Ghats, India and its sustainable utilization through *in vitro* technology. *In Vitro Cellular & Developmental Biology: Plant, 47*, 110–122.

Krishnaveni, A., & Thaakur, S. R. (2008). Pharmacognostical and preliminary phytochemical studies of *Sapindus trifoliatus* Vahl. *Ethnobotanical Leaflets, 12*, 820–826.

Kumar, S., Mangal, M., Dhawan, A. K., & Singh, N. (2011). Assessment of genetic fidelity of micropropagated plants of *Simmondsia chinensis* (link) Schneider using RAPD and ISSR markers. *Acta Physiologae Plantarum, 33*, 2541–2545.

Kumar, V., Murthy, K. N., Bhamid, S., Sudha, C. G., & Ravishankar, G. A. (2005). Genetically modified hairy roots of *Withania somnifera* Dunal: A potent source of rejuvenating principles. *Rejuvenation Research, 8*, 37–45.

Kumar, V., Moyo, M., Gruz, J., Subrtova, M., & Staden, J. V. (2015). Phenolic acid profiles and antioxidant potential of *Pelargonium sidoides* callus cultures. *Industrial Crops and Products, 77*, 402–408.

Kumar, M. N., Govindh, B., & Annapurna, N. (2017). Green synthesis and characterization of platinum nanoparticles using *Sapindus mukorossi* Gaertn. Fruit pericarp. *Asian Journal of Chemistry, 29*, 2541–2544.

Kuo, Y., Huang, H. C., Kuo, L. M. Y., Hsu, Y. W., Lee, K. H., Chang, F. R., & Wu, Y. C. (2005). New demmarane-type saponins from galls of *S. mukorossi*. *Journal of Agricultural and Food Chemistry, 53*, 4722–4727.

Kurczyjska, E. U., Gaj, M. D., Ujczak, A., & Mazur, E. (2007). Histological analysis of direct somatic embryogenesis in Arabidopsis thaliana (L.) Heynh. *Planta, 226*, 619–628.

Lai, H. Y., & Lim, Y. Y. (2011). Evaluation of antioxidant activities of the methanolic extracts of selected ferns in Malaysia. *International Journal of Environmental Science and Development, 2*, 442–447.

Lai, F. M., & McKersie, B. D. (1993). Effect of nutrition on maturation of alfalfa (Medicago sativa L.) somatic embryos. *Plant Sciences, 91*, 87–95.

Langhansova, L., Konradova, H., & Vanek, T. (2004). Polyethylene glycol and abscisic acid improve maturation and regeneration of Panax ginseng somatic embryos. *Plant Cell Reports, 22*, 725–730.

Larkin, P. J., & Scowcroft, W. R. (1981). Somaclonal variation – A novel source of variability from cell cultures for plant improvement. *Theoretical and Applied Genetics, 60*, 197–214.

Larkin, P. J., Banks, P. M., Bhati, R., Brettell, R. I. S., Davies, P. A., Ryan, S. A., Scowcroft, W. R., Spindler, L. H., & Tanner, G. J. (1989). From somatic variation to variant plants: Mechanism and applications. *Genome, 31*, 705–711.

Lee, M. H., Jeong, J. H., Seo, J. W., Shin, C. G., Kim, Y. S., In, J. G., Yang, D. C., Yi, J. S., & Choi, Y. E. (2004). Enhanced triterpene and phytosterol biosynthesis in *Panax ginseng* overexpressing squalene synthase gene. *Plant & Cell Physiology, 45*, 976–984.

Lee, W. S., Kim, J. R., Han, J. M., Jang, K. C., Sok, D. E., & Jeong, T. S. (2006). Antioxidant activities of abietane diterpenoids isolated from *Torreya nucifera* leaves. *Journal of Agricultural and Food Chemistry, 54*, 5369–5374.

Lema-Ruminska, J., Goncerzewicz, K., & Gabriel, M. (2013). Infuence of abscisic acid and sucrose on somatic embryogenesis in Cactus *Copiapoa tenuissima* Ritt. forma *mostruosa*. *The Scientific World Journal, 2013*, 713985. https://doi.org/10.1155/2013/513985.

Lemos, T. L. G., Mendes, A. L., Sousa, M. P., & Braz-Filho, R. (1992). New saponin from Sapindus saponaria. *Fitoterapia, 93*, 515–517.

Lemos, T. L. G., Sousa, M. P., Mendes, A. L., & Braz-Filho, R. (1994). Saponin from Sapindus saponaria. *Fitoterapia, 95*, 557–558.

Linde, H. (1979). Uberinhaltsstoffe der perikarpein von *S. mukorossi* gaertn. *Archives of Pharmacal Research, 312*, 416–425.

Linossier, L., Veisseire, P., Cailloux, F., & Coudret, A. (1997). Effects of abscisic acid and high concentrations of PEG on Hevea brasiliensis somatic embryos development. *Plant Science, 124*, 183–191.

Litz, R. E., & Gray, D. J. (1992). Organogenesis and somatic embryogenesis. In F. A. Hammerschlag & R. E. Litz (Eds.), *Biotechnology of pernnial fruit crops* (pp. 3–34). Wallingford: CAB International.

Litz, R. E., & Gray, D. J. (1995). Somatic embryogenesis for agricultural improvement. *World Journal of Microbiology and Biotechnology, 11*, 416–425.

Litz, R. E., & Jaiswal, V. S. (1991). Micropropagation of tropical and subtropical fruits. In P. C. Debergh & R. H. Zimmerman (Eds.), *Micropropagation* (pp. 247–263). Dordrecht: Kluwer Academic Publishers.

Liu, C. Z., Murch, S. J., El Demerdash, M., & Saxena, P. K. (2004). *Artemisia judaica* L: Micropropagation and antioxidant activity. *Journal of Biotechnology, 110*, 63–71.

Lakshmanan, V., Venkataramareddy, S. R., & Neelwarne, B. (2007). Molecular analysis of genetic stability in long-term micropropagated shoots of banana using RAPD and ISSR markers. *Electronic Journal of Biotechnology, 10*, 1–8.

Lattoo, S. K., Bamotra, S., Sapru, D. R., Khan, S., & Dha, A. K. (2006). Rapid plant regeneration and analysis of genetic fidelity of Chlorophytum arundinaceum baker – An endangered medicinal herb. *Plant Cell Reports, 25*, 499–506.

Lugato, D., Simao, M. J., Garcia, R., Mansur, E., & Pacheco, G. (2014). Determination of antioxidant activity and phenolic content of extracts from in vivo plants and in vitro materials of *Passiflora alata* Curtis. *Plant Cell, Tissue and Organ Culture, 118*, 339–346.

Mahabusarakam, W., Towers, G. H. N., Tuntiwachwuttikul, P., & Wiriyachitra, P. (1990). Pesticidal triterpenoid saponins of the pericarps of S. emargiaus. *Journal of the Science Society of Thailand, 16*, 187.

Mahmood, A., Asif, I. M., Alam, M. S., Islam, M. S., Eti, S. A., & Hossain, F. (2013). Extraction and characterization of oils from Sapindus trifoliatus Linn seeds of different origin of Bangladesh. *Merit Research Journal of Environmental Science and Toxicology, 1*, 99–104.

Maisarah, A. M., Nurul-Amira, B., Asmah, R., & Fauziah, O. (2013). Antioxidant analysis of different parts of *Carica papaya*. *International Food Research Journal, 20*, 1043–1048.

Makarov, V. V., Love, A. J., Sinitsyna, O. V., Makarova, S. S., Yaminsky, I. V., Taliansky, M. E., et al. (2014). Green nanotechnologies: Synthesis of metal nanoparticles using plants. *Acta Naturae, 6*, 35.

Mali, R. R., Kumar, A., Singh, A. K., & Talwar, A. (2010). Formulation of herbal shampoos from *Asparagus racemosus, Acacia concin, Sapindus mukorossi*. *International Journal of Pharmaceutical Sciences Review and Research, 4*, 39–44.

Malik, P., Shankar, R., Malik, V., Sharma, N., & Mukherjee, T. K. (2014). Green chemistry based benign routes for nanoparticle synthesis. *The Journal of Nanoparticle, 2014*, 1–14. https://doi.org/10.1155/2014/302429.

Mallikarjuna, K., & Rajendrudu, G. (2007). High frequency in vitro propagation of Holarrhena antidysenterica from nodal buds of mature tree. *Biologia Plantarum, 51*, 525–529.

Martin, K. P. (2002). Rapid propagation of *Holostemma ada-kodien* Schult., a rare medicinal plant through axillary bud multiplication and indirect organogenesis. *Plant Cell Reports, 21*, 112–117.

Martins, M., Sarmento, D., & Oliveira, M. M. (2004). Genetic stability of micropropagated almond plantlets, as assessed by RAPD and ISSR markers. *Plant Cell Reports, 23*, 492–496.

Martin, K. P., Pachathudikandi, S., Zhang, C. L., Slater, A., & Madassery, J. (2006). RAPD analysis of a variant of banana (Musa sp.) cv. Grande naine and its propagation via shoot tip culture. *In vitro Cellular and Developmental Biology-Plant, 42*, 188–192.

Martínez, M. T., Vieitez, A. M., & Corredoira, E. (2015). Improved secondary embryo production in *Quercus alba* and *Quercus rubra* by activated charcoal, silver thiosulphate and sucrose: Influence of embryogenic explant used for subculture. *Plant Cell Tissue and Organ Culture, 121*, 531–546.

Martinelli, L., & Mandolino, G. (2009). Strategies for effective somatic embryogenesis in grapevine (Vitis spp.): An appraisal. In K. A. Roubelakis-Angelakis (Ed.), *Grapevine molecular physiology and biotechnology* (p. 612). Dordrecht: Springer.

Martinez, F. D. (2007). Genes, environments, development and asthma: A reappraisal. *European Respiratory Journal, 29*, 179–184.

Martinez, M. J., Betancourt, J., Alonso-Gonzalez, N., & Jauregui, A. (1996). Screening of some Cuban medicinal plants for antimicrobial activity. *Journal of Ethnopharmacology, 52*, 171–174.

Maryanti, E., Damayanti, D., Gustian, I., & Yudha, S. (2014). Synthesis of ZnO nanoparticles by hydrothermal method in aqueous rinds extracts of Sapindus rarak DC. *Materials Letters, 118*, 96–98.

Matkowski, A. (2008). Plant in vitro culture for the production of antioxidants — A review. *Biotechnology Advances, 26*, 548–560.

McComb, J. A., & Bennett, I. J. (1982). Vegetative propagation of Eucalyptus using tissue culture and its application to forest improvement in western Australia. In A. Fujiwara (Ed.), *Plant tissue culture* (pp. 721–722). Tokyo: Maruzen.

McComb, J. A., & Bennett, I. J. (1986). Eucalyptus (*Eucalyptus* spp.). In Y. P. S. Bajaj (Ed.), *Biotechnology in agriculture and forestry, Vol 1* (Trees 1) (pp. 340–362). Berlin/Hiedelberg: Springer.

McDonald, S., Prenzler, P. D., Autolovich, M., & Robards, K. (2001). Phenolic content and antioxidant activity of olive extracts. *Food Chemistry, 73*, 73–84.

Meena, V. N., Rajakohila, M., Syndia, L. A. M., Prasad, P. N., & Ariharan, V. N. (2012). Multifacetious uses of soapnut tree – A mini review. *Research Journal of Pharmaceutical, Biological and Chemical Sciences, 3*, 420–424.

Mehata, M. S. (2015). Enhancement of charge transfer and quenching of photoluminescence of capped CdS quantum dots. *Scientific Reports, 5*, 12056. https://doi.org/10.1038/srep12056.

Merkle, S. A. (1995). Strategies for dealing with limitations of somatic embryogenesis in hardwood trees. *Plant Tissue Culture and Biotechnology, 1*, 112–121.

Merkle, S. A., & Dean, J. F. D. (2000). Forest biotechnology. *Current Opinion in Biotechnology, 11*, 298–302.

Michler, C. H., & Bauer, E. O. (1991). High frequency somatic embryogenesis from leaf tissue of Populus spp. *Plant Science, 77*, 111–118.

Miller, H. E. (1971). A simplified method for the evaluation of antioxidants. *Journal of the American Oil Chemists Society, 48*, 91–91.

Mittal, A. K., Chisti, Y., & Banerjee, U. C. (2013). Synthesis of metallic nanoparticles using plant extracts. *Biotechnology Advances, 31*, 346–356.

Moon, H. K., Kim, Y. W., Lee, J. S., & Choi, Y. E. (2005). Micropropagation of Kalopanax Pictus tree via somatic embryogenesis. *In Vitro Cellular and Developmental Biol.—Plant, 41*, 303–306.

Moreira, A. L., Scariot, D. B., Pelegrini, B. L., Pessini, G. L., Ueda-Nakamura, T., Nakamura, C. V., & Ferreira, I. C. P. (2017). Acyclic Sesquiterpenes from the fruit pericarp of *Sapindus saponaria* induce ultrastructural alterations and cell death in *Leishmania amazonensis*. *Evidence-based Complementary and Alternative Medicine, 2017*, 5620693., 11 pp. https://doi.org/10.1155/2017/5620693.

Morikawa, T., Xie, Y., Asao, Y., Okamoto, M., Yamashita, C., Muraoka, O., Matsuda, H., Pongpiriyadacha, Y., Yuan, D., & Yoshikawa, M. (2009). Oleanane type triterpene oligoglycosides with pancreatic lipase inhibitory activity from the pericarps of S. rarak. *Phytochemistry, 70*, 1166–1172.

Munson, R. H. (1984). Germination of western soapberry as affected by scarification and stratification. *Horticultural Science, 19*, 712–713.

Murashige, T. (1974). Plant propagation through tissue culture. *Annual Review of Plant Physiology, 25*, 135–166.

Murashige, T., & Skoog, F. A. (1962). Revised medium for rapid growth and bioassays with tobacco tissue cultures. *Physiologia Plantarum, 15*, 473–497.

Nabavi, S. M., Ebrahimzadeh, M. A., Nabavi, S. F., Hamidinia, A., & Bekhradnia, A. R. (2008). Determination of antioxidant activity, phenol and flavonoids content of *Parrotia persica* Mey. *Pharmacology Online, 2*, 560–567.

Nagvani, V., Madhavi, Y., Rao, D. B., Rao, P. K., & Rao, T. R. (2010). Free radical scavenging activity and qualitative analysis of polyphenols by RP-HPLC inn the flowers of *Couroupita guianensis* Aubl. *The Electronic Journal of Environmental, Agricultural and Food Chemistry, 9*, 1471–1484.

Nakayama, K., Fujino, H., Kasai, R., Mitoma, Y., Yata, N., & Tanaka, O. (1986). Solubilizing properties of saponins from *S. mukorossi* Gaertn. *Chemical & Pharmaceutical Bulletin, 34*, 3279–3283.

Narayanan, K. B., & Sakthivel, N. (2010). Biological synthesis of metal nanoparticles by microbes. *Advances in Colloid and Interface Science, 156*, 1–13.

Natesh, S. (1999). Conservation of medicinal and aromatic plants in India – An overview. In S. Natesh, A. Osman, & A. K. Azizol (Eds.), *Medicinal and aromatic plants: Strategies and technologies for conservation* (pp. 1–11). Kuala Lumpur: Forest Research Institute.

Ni, W., Hua, Y., Teng, R. W., Kong, Y. C., & Chen, C. X. (2004). New tirucallane-type triterpenoid saponins from Sapindus mukorossi. *Journal of Asian Natural Product Research, 6*, 205–209.

Ni, W., Hua, Y., Liu, H. Y., Teng, R. W., Kong, Y. C., Hu, X. Y., & Chen, C. X. (2006). Tirucallane-type triterpenoid saponins from the roots of *S. mukorossi*. *Chemical and Pharmaceutical Bulletin, 54*, 1443–1446.

Nianhe, X., & Gadek, P. A. (2012). Sapindus rarak, Flora of China. Missouri Botanical Garden, St. Louis, MO & Harvard University Herbaria, Cambridge, MA. Retrieved June 5, 2012.

Nomura, K., & Komamine, A. (1995). Physiological and biochemical aspects of somatic embryogenesis. In T. A. Thorpe (Ed.), *In vitro embryogenesis in plants* (pp. 249–265). Dordrecht: Kluwer Academic Publisher.

Ochiai, T., Ohno, S., Soeda, S., Tanaka, H., Shoyama, Y., & Shimeno, H. (2004). Crocin prevents the death of rat pheochromyctoma (PC-12) cells by its antioxidant effects stronger than those of alpha-tocopherol. *Neuroscience Letters, 362*, 61–64.

Ohkawa, H., Ohishi, N., & Yagi, K. (1979). Assay for lipid peroxides in animal tissues by thiobarbituric acid reaction. *Analytical Biochemistry, 95*, 351–358.

Oleszek, W. A. (2000). Saponins. In *Natural foods antimicrobial systems*. Boca Raton: CRC Press, LLC.

Oropeza, M., Guevara, P., García, E., et al. (1995). Identification of sugarcane (Saccharum spp.) somaclonal variants using randomly amplified polymorphic DNA (RAPDs). *Plant Molecular Biology Reports, 13*, 169–180.

Orwa, C., Mutua, A., Kindt, R., Jamnadass, R., & Simons, A. (2009). *Agroforestree database: A tree reference and selection guide version 4.0.* http://www.worldagroforestry.org/af/treedb/

Pal, R., Girhepunjem, K., Shrivastav, N., Hussain, M. M., & Thirumoorthy. (2011). Antioxidant and free radical scavenging activity of ethanolic extract of *Morinda citrifolia*. *Annals of Biological Research, 2*, 127–131.

Pandey, S. (2003). *Regeneration potential of some somatic tissues of Terminalia arjuna Roxb. and Sapindus trifoliatus L.* PhD thesis, Banaras Hindu University, Varanasi, India.

Pandey, V. K., & Sharma, A. K. (1986). Evaluation of 'Vathari gugul and Nadivaspa-Sweda' in the management of rheumatic diseases. *Rheumatism, 22*, 1–6.

Pandey, S., Singh, M., Jaiswal, U., & Jaiswal, V. S. (2006). Shoot initiation and multiplication from a mature tree of *Terminalia arjuna* Roxb. *In Vitro Cellular & Developmental Biology: Plant, 42*, 389–393.

Pandey, N., Chaurasia, J. K., Tiwari, O. P., & Tripathi, Y. B. (2007). Antioxidant properties of different fractions of tubers from *Pueraria tuberosa* Linn. *Food Chemistry, 105*, 219–222.

Parashar, V., Parashar, R., Sharma, B., & Pandey, A. (2009). Parthenium leaf extract mediated synthesis of silver nanoparticles: A novel approach towards weed utilization. *Digest Journal of Nanomaterials and Biostructures, 4*, 45–50.

Parganiha, R., Mishra, J., Deshmukh, V. S., Muley, B. P., Sawarkar, H. A., & Kashyap, P. (2011). In- vitro anti- asthmatic activity of fruit extract of *Sapindus mukorossi*. *The International Journal of Herbal Medicine, 1*, 11–14.

Parsaeimehr, A., Sargsyan, E., & Javidni, K. (2010). A comparative study of the antibacterial, antifungal and antioxidant activity and total content of phenolic compounds of cell cultures and wild plants of three endemic species of *Ephedra*. *Molecules, 15*, 1668–1678.

Pavlov, A., Georgiev, V., & Ilieva, M. (2005). Betalain biosynthesis by red beet (Beta vulgaris L.) hairy root culture. *Process Biochemistry, 40*, 1531–1533.

Pavlov, A., & Bley, T. (2006). Betalains biosynthesis by Beta vulgaris L. hairy root culture in a temporary immersion cultivation system. *Process Biochemistry, 41*, 848–852.

Pelegrini, D. D., Tsuzuki, J. K., Amado, C. A. B., Cortez, D. A. G., & Ferreira, I. C. P. (2008). Biological activity and isolated compounds in *Sapindus saponaria* L. and other plants of the genus *Sapindus*. *Latin American Journal of Pharmacy, 27*, 922–927.

Petersen, M., & Simmonds, M. S. (2003). Rosmarinic acid. *Phytochemistry, 62*, 121–125.

Phillips, R. L., Kaeppler, S. M., & Olhoft, P. (1994). Genetic instability of plant tissue cultures: Breakdown of normal controls. *Proceeding of National Academy of Sciences-USA, 91*, 5222–5226.

Phillips, D. R., Rasbery, J. M., Bartel, B., & Matsuda, S. P. T. (2006). Biosynthetic diversity in plant triterpene cyclization. *Current Opinion in Plant Biology, 9*, 305–314.

Philomina, N. S. (2010). Somatic embryogenesis from leaf explants of soapnut (*Sapindus mukorossi* Gaertn.). *Indian Journal of Biotechnology, 9*, 336–337.

Philomina, N. S., & Rao, J. V. S. (2000). Micropropagation of *Sapindus mukorossi* Gaertn. *The Indian Journal of Experimental Biology, 38*, 621–624.

Porchezhiyana, V., & Noorjahan, S. E. (2016). Fabrication of a versatile chitosan nanocomposite hydrogel impregnated with biosynthesized silver nanoparticles using *Sapindus Mukorossi:* Characterization and applications. *RSC Advances, 6*(98), 95564–95573. https://doi.org/10.1039/C6RA19546F.

Prabhakar, T., Bhogavalli, P. K., Bathula, N. R. V., Gangula, S. R., & Chenreddy, S. K. (2012). *In vitro* studies on antimicrobial screening of leaf extracts of *Sapindus Saponaria* against common dental pathogens. *Plant Sciences Feed, 2*, 15–18.

Pradhan, D. (2014). Pharmacological effect of some fractions obtained from *Sapindus trifoliatus* acting as an antioxidant and against mammary cell proliferation. *African Journal of Pharmacy and Pharmacology, 8*, 455–463.

Pradhan, D., & Tripathy, G. (2014). Effect of *Sapindus trifoliatus* leaf extracts on human breast cancer cell lines. *Research Journal of Pharmaceutical, Biological and Chemical Sciences, 5*, 845–852.

Prakash, M. G., & Gurumurthi, K. (2010). Effects of type of explant and age, plant growth regulators and medium strength on somatic embryogenesis and plant regeneration in *Eucalyptus camaldulensis*. *Plant Cell, Tissue and Organ Culture, 100*, 13–20.

Prinz, A. (1991). Ethnomedicine – History of medicine. Sciences of medical thinking and behavior of man. *Wiener Klinische Wochenschrift, 103*, 481–483.

Purohit, S. D., & Kukda, G. (2004). Micropropagation of an adult tree, Wrightia tinctoria. *Indian Journal of Biotechnology, 3*, 216–220.

Quatrano, R. S. (1989). The role of hormones during seed development. In R. D. Davies (Ed.), *Plant hormones and their role in plant growth and development* (pp. 494–514). Dordrecht: Martinus Nijhoff Publishers.

Raghavan, V. (2000). *Developmental biology of flowering plants*. New York: Springer-Verlag.

Raghuvanshi, P., Bagga, R., Malhotra, D., Gopalan, S., & Talwar, G. P. (2001). Spermicidal & contraceptive properties of Praneem polyherbal pessary. *The Indian Journal of Medical Research, 113*, 135–141.

Rahman, S. S., Rahman, M. M., & Begum, S. A. (2007). Investigation of *Sapindus mukorossi*extracts for repellency, insecticidal activity and plant growth regulatory effect. *Journal of Applied Sciences Research, 3*, 95–101.

Rajeswari, V., & Paliwal, K. (2006). *In vitro* propagation of *Albizia odoratissima* L.F. (Benth.) from cotyledonary node and leaf nodal explants. *In Vitro Cellular & Developmental Biology: Plant, 42*, 399–404.

Ramgopal, M., Saisushma, C. H., Attitalla, I. H., & Alhasin, A. M. (2011). A facile green synthesis of silver nanoparticles using soap nuts. *Research Journal of Microbiology, 6*, 432–438.

Rani, V., Parida, A., & Raina, S. N. (1995). Random amplified polymorphic DNA (RAPD) markers for genetic analysis in micropropagated plants of Populus deltoidecs marsh. *Plant Cell Reports, 14*, 459–462.

Rani, V., & Raina, S. N. (2000). Genetic fidelity of organized meristem derived micropropagated plants: A critical reappraisal. *In Vitro Cellular and Developmental Biology – Plant, 36*, 319–330.

Rao, P. S. (1965). In vitro induction of embryonal proliferation in Santalum album L. *Phytomorphology, 15*, 175–179.

Rao, G. H. J., & Lakshmi, P. (2012). *Sapindus trifoliatus*: A review. *International Journal of Pharmacy and Technology, 4*, 2201–2204.

Ray, T., Dutta, I., Saha, P., Das, S., & Roy, S. C. (2006). Genetic stability of three economically important micropropagated banana (Musa spp.) cultivars of lower indo-Gangetic plains, as assessed by RAPD and ISSR markers. *Plant Cell Tissue and Organ Culture, 85*, 11–21.

Rashed, K. N., Ciric, A., Glamoclija, J., Calhelha, R. C., Ferreira, I. C. F. R., & Sokovic, M. (2013). Antimicrobial activity, growth inhibition of human tumour cell lines, and phytochemical characterization of the hydromethanolic extract obtained from *Sapindus saponaria* L. aerial parts. *BioMed Research International, 2013*, 659183., 9 pp. https://doi.org/10.1155/2013/659183.

Ratnesh, R. K., & Mehata, M. S. (2015). Controlled synthesis and optical properties of tunable CdSe quantum dots and effect of pH. *AIP Advances, 5*, 097114. https://doi.org/10.1063/1.4930586.

Ravindra, P. V., & Narayan, M. S. (2003). Antioxidant activity of the anthocyanin from carrot (*Daucus carota*) callus culture. *International Journal of Food Sciences and Nutrition, 54*, 349–355.

Ravishankara, M. N., Shrivastava, N., Padh, H., & Rajani, M. (2002). Evaluation of antioxidant properties of root bark of *Hemidesmus indicus* R. Br. (Anantmul). *Phytomedicine, 9*, 153–160.

Ray, A., & Bhattacharya, S. (2010). Storage and conversion of *Eclipta alba* synseeds and RAPD analysis of the converted plantlets. *Biologia Plantarum, 54*, 547–550.

Reddy, M. P., Sarla, N., & Siddiq, E. A. (2002). Inter simple sequence repeat (ISSR) polymorphism and its application in plant breeding. *Euphytica, 128*, 9–17.

Reddy, V., Torati, R. S., Oh, S., & Kim, C. G. (2013). Biosynthesis of gold nanoparticles assisted by *Sapindus mukorossi* Gaertn. Fruit pericarp and their catalytic application for the reduction of p-nitroaniline. *Industrial and Engineering Chemistry Research, 52*(2), 556–564. https://doi.org/10.1021/ie302037c.

Ribeiro, A., Zani, C. L., Alves, T. M. A., Mendes, N. M., Hamburger, M., & Hostettmann, K. (1995). Molluscicidal saponins from the pericarp of Sapindus saponaria. *International Journal of Pharmacology, 33*, 177–180.

Reinert, J. (1958). Untersuchungen fiber die Morphogenese an Gewebekulturen. *Berichte der Deutschen Botanischen Gesellschaft, 71*, 15.

Risterucci, A. M., Duval, M. F., Rohde, W., & Billotte, N. (2005). Isolation and characterization of microsatellite loci from Psidium guajava. *Molecular Ecology Notes, 5*, 745–748.

Robichaud, R. L., Lessard, V. C., & Merkle, S. A. (2004). Treatments affecting maturation and germination of American chestnut somatic embryos. *Journal of Plant Physiology, 161*, 957–969.

Rocha, D. I., Vieira, L. M., Tanaka, F. A., Silva, L. C., & Otoni, W. C. (2012). Somatic embryogenesis of a wild passion fruit species Passiflora cincinnata masters: Histocytological and histochemical evidences. *Protoplasma, 249*, 747–758.

Ruch, R. J., Cheng, S. J., & Klaunig, J. E. (1989). Prevention of cytotoxicity and inhibition of intracellular communication by antioxidant catechins isolated from Chinese green tea. *Carcinogenesis, 10*, 1003–1008.

Rugini, E., & Caricato, G. (1995). Somatic embryogenesis and plant recovery from mature tissues of olive cultivars (Olea europaea L.) "Canino and Moraiolo". *Plant Cell Reports, 14*, 257–260.

Sharma, M. M., Verma, R. N., Singh, A., & Bata, A. (2014). Assessment of clonal fidelity of Tylophora indica (Burm. F.) Merrill "in vitro" plantlets by ISSR molecular markers. *Springer Plus, 3*, 1–9.

Sahoo, P. K., Pradhan, D., & Behra, P. (2010a). Pharmacological interventions of some potential herbal extracts targeted at adenosine receptors in diabetic neuropathic pain. *Journal of Pharmacy Research, 3*, 1798–1802.

Sahoo, P. K., Padhy, K. M., Pradhan, D., Tripathy, G., Bhoi, R., Pattanayak, S., & Sahoo, S. K. (2010b). Anti-diabetic and antioxidant activity of ethanolic extract of *sapindus trifoliatus* linn. *International Journal of Pharma and Bio Sciences, 1*, 1–8.

Sagare, A. P., Lee, Y. L., Lin, T. C., Chen, C. C., & Tsay, H. S. (2000). Cytokinin induced somatic embryogenesis and plant regeneration in Corydalis yanhusuo (Fumariaceae)—A medicinal plant. *Plant Science, 160*, 139–147.

Sarmast, M. K., Salehi, H., Ramezani, A., Abolimoghadam, A. A., Niazi, A., & Khosh-Khui, M. (2012). RAPD fingerprint to appraise the genetic fidelity of in vitro propagated Araucaria excels R. Br. Var. glauca plantlets. *Molecular Biotechnology, 50*, 181–188.

Saxena, D., Pal, R., Dwivedi, A. K., & Singh, S. (2004). Characterization of sapindosides in *Sapindus mukorossi* saponin (Reetha saponin) and quantitative determination of sapindoside B. *Journal of Scientific and Industrial Research, 63*, 181–186.

Schuster, B. G. (2002). Development of antimalarial agents and drugs for parasitic infections based on leads from traditional medicine: The Walter Reed experience. *Advances in Phytomedicine, 1*, 163–171.

Sengupta, A., Basu, S. P., & Saha, S. (1975). Triglyceride composition of *S. mukorossi* seed oil. *Lipids, 10*, 33–40.

Sergeev, G. B., & Shabatina, T. I. (2008). Cryochemistry of nanometals. *Colloids and Surfaces A: Physicochemical and Engineering Aspects, 313*, 18–22.

Sharma, N., & Patni, V. (2013). Comparative analysis of total flavonoids, quercetin content and antioxidant activity of in vivo and in vitro plant parts of *Grewia asiatica* Mast. *International Journal of Pharmacy and Pharmaceutical Sciences, 5*, 464–469.

Sharma, A., Sati, S. C., Sati, O. P., Sati, D., & Kothiyal, S. K. (2011). Chemical constituents and bioactivities of genus *Sapindus*. *International Journal of Research in Ayurveda and Pharmacy, 2*, 403–409.

Sharma, A., Satish, C. S., Sati, O. P., Sati, M. D., Kothiyal, S. K., Semwal, D. K., & Mehta, A. (2013). A new triterpenoid saponin and antimicrobial activity of ethanolic extract from *Sapindus mukorossi* Gaertn. *Journal of Chemistry, 2013*, 613190., 5 pp. https://doi.org/10.1155/2013/613190.

Sharp, W. R., Sondhal, M. R., Caldas, L. S., & Maraffa, S. B. (1980). The physiology of in vitro asexual embryogenesis. *Horticultural Reviews, 2*, 268–301.

Shiau, I. L., Shih, T. L., Wang, Y. N., Chen, H. T., Haw–Farn Lan, H. F., Lin, H. C., Yang, B. Y., Ko, C. H., & Murase, Y. (2009). Quantification for Saponin from a soapberry (Sapindus mukorossi Gaertn) in cleaning products by a chromatographic and two colorimetric assays. *Journal of the Faculty of Agriculture Kyushu University, 54*, 215–221. https://www.google.co.in/search?q=sapindus+rarak&sxsrf=ACYBGNQxi0ZH11NeFz3ajDFWmDsO-tMYgA:1569327331674&source=lnms&tbm=isch&sa=X&ved=0ahUKEwir3Y_xt-nkAhXRF3IKHRWZB4EQ_AUIESgB&biw=1366&bih=654#imgrc=LGUODMH2JrbEKM; https://alchetron.com/Sapindus-rarak

Shu, Q. Y., Liu, G. S., Qi, D. M., Chu, C. C., Liu, J., & Li, H. J. (2003). An effectivemethod for axillary bud culture and RAPD analysis of cloned plants in tetraploid black locust. *Plant Cell Reports, 22*, 175–180.

Shukla, S., Shukla, S. K., & Mishra, S. K. (2009). *In vitro* plant regeneration from seedling explants of *Stereospermum personatum* D.C.: A medicinal tree. *Trees-Structure and Function, 23*, 409–413.

Shyamkumar, B., Anjaneyulu, C., & Giri, C. C. (2003). Multiple shoot induction from cotyledonary node explants of *Terminalia chebula* Retz: A tree of medicinal importance. *Biologia Plantarum, 47*, 585–588.

Siddique, I., & Anis, M. (2007). Rapid micropropagation of *Ocimum basilicum*using shoot tip explants pre-cultured in thidiazuron supplemented liquid medium. *Biologia Plantarum, 51*, 787–790.

Slik, F. (2012). Sapindus rarak. Plants of Southeast Asia. Retrieved June 5, 2012, from https://en.wikipedia.org/wiki/Sapindus_rarak

Silva, M. L. D., Marcussi, S., Fernandes, R., Pereira, P. S., Januario, A. H., Franca, S. C., Silva, S. L. D., Soares, A. M., & Lourenco, M. V. (2012). Anti-snake venom activities of extracts and fractions from callus cultures of Sapindus saponaria. *Pharmaceutical Biology, 50*, 366–375.

Singh, D. K., & Agarwal, R. A. (1983). In-vivo and in-vitro studies on synergism with anticholinesterase pesticides in the snail *Lymnaea acuminate. Arch Environ Contain Toxicol, 12*, 483–487.

Singh, N., Kaur, A., & Yadav, K. (2010). A reliable in vitro protocol for rapid mass propagation of Sapindus mukorossi Gaertn. *Nature & Science, 8*, 41–47.

Singh, R., & Kumari, N. (2015). Comparative determination of phytochemicals and antioxidant activity from leaf and fruit of *Sapindus mukorossi* Gaertn. – A valuable medicinal tree. *Industrial Crops and Products, 73*, 1–8.

Singh, K., Panghal, M., Kadyan, S., & Yadav, J. P. (2014). Evaluation of antimicrobial activity of synthesized silver nanoparticles using *Phyllanthus amarus* and *Tinospora cordifolia* medicinal plants. *Journal of Nanomedicine and Nanotechnology, 5*, 250. https://doi.org/10.4172/2157-7439.1000250.

Singh, R., Rai, M. K., & Kumari, N. (2015). Somatic embryogenesis and plant regeneration in *Sapindusn mukorossi* Gaertn. from leaf-derived callus induced with 6-Benzylaminopurine. *Applied Biochemistry and Biotechnology, 177*, 498–510.

Singh, R., Kashyap, S. P., Kumari, N., & Singh, M. (2016a). Regeneration of soapnut tree through somatic embryogenesis and assessment of genetic fidelity through ISSR and RAPD markers. *Physiology and Molecular Biology of Plants, 22*, 381–389.

Singh, R., Kumari, N., & Nath, G. (2016b). Antimicrobial efficacy of callus and in vitro leaf extracts of *Sapindus Mukorossi* Gaertn. Against pathogenic microbes. *M J Pharmaceuticals, 1*, 1–4.

Singh, R., Kumari, N., & Nath, G. (2016c). Free radicals scavenging activity and antimicrobial potential of leaf and fruit extractsof *Sapindus mukorossi* Gaertn. against clinical pathogen. *International Journal of Phytomedicine, 8*, 22–28.

Sohn, J. S., Kwon, Y. W., Jin, J. I., & Jo, B. W. (2011). DNA-templated preparation of gold nanoparticles. *Molecules, 16*, 8143–8151.

Sommer, H. E., Brown, C. L., & Kormanik, P. P. (1975). Differentiation of plantlets in long leaf pine (*Pinus palustris* Mill) tissue cultured *in vitro*. *Botanical Gazette, 135*, 196–200.

Soobrattee, M. A., Neergheen, V. S., Luximon-Ramma, A., Aruoma, O. I., & Bahorun, T. (2005). Phenolics as potential antioxidant therapeutic agents: Mechanisms and actions. *Mutation Research, 579*, 200–213.

Sreekala-devi, R., Radhamany, P. M., & Gayathri Devi, V. (2013). Investigation of the antioxidant principles from *Psilanthus travancorensis* (WT. ARN.) Leroy – An unexplored taxon of Rubiaceae. *International Journal of Pharmaceutical Sciences, 5*, 13–17.

Srikanth, J., & Muralidharan, P. (2009a). Anti-hyperglycemic and anti-diabetic activity of leaves extracts of *Sapindus emarginatus* Vahl. *Asian Biomedicine, 3*, 313–318.

Srikanth, J., & Muralidharan, P. (2009b). CNS activity of the methanolic extracts of *Sapindus emarginatus* Vahl in experimental models. *Journal of Scientific Research, 1*, 583–593.

Stasolla, C., & Yeung, E. C. (2003). Recent advances in conifer somatic embryogenesis: Improving somatic embryo quality. *Plant Cell, Tissue and Organ Culture, 74*, 15–35.

Steward, F. C., Mapes, M. O., & Mears, K. (1958). Growth and organised development of cultured cells: II. Organisation in cultured grown from freely suspended cells. *American Journal of Botany, 45*, 705–707.

Stiffness, M., & Douros, J. (1982). Current status of the NCL plants and animals product programme. *Journal of Natural Products, 45*, 1–45.

Stoffels, K. (2008). Soap nut saponins create powerful natural surfactant. *Personal Care Magazine*. Jeen International Corporation.

Subha, B., Prasath, P. V., Abinaya, R., Kavitha, R. J., & Ravichandran, K. (2015). Synthesis and characterization of nano-hydroxyapatite using Sapindus Mukorossi extract. *AIP Conference Proceedings. Solid State Physics, 1665*, 010001. https://doi.org/10.1063/1.4917768.

Suhagia, B. N., Rathod, I. S., & Sindhu, S. (2011). *Sapindus mukorossi* (Areetha): An overview. *International Journal of Pharmaceutical Sciences and Research, 2*, 1905–1913.

Swartz, H. W., Galletta, G. J., & Zimmerman, R. H. (1991). Field performance and phenotypic stability of tissue culture propagated strawberries. *Journal of the American Society for Horticultural Science, 106*, 667–673.

Tadhani, M. B., Patel, V. H., & Subhash, R. (2007). In vitro antioxidant activities of *Stevia rebaudiana* leaves and callus. *Journal of Food Composition and Analysis, 20*, 323–329.

Takagi, K., Park, E. H., & Kato, H. (1980). Anti-inflammatory activities of hederagenin and crude saponin isolated from *Sapindus mukorossi* Gaertn. *Chemical & Pharmaceutical Bulletin (Tokyo), 28*, 1183–1188.

Tanaka, O., Tamura, Y., Masuda, H., & Mizutani, K. (1996). In G. R. Waller & K. Yamasaki (Eds.), *Application of saponins in food and cosmetics: Saponins of Mohova Yucca and Sapindus mukorossi Gaertn, saponins used in food and agriculture* (pp. 1–11). New York: Plenum Press.

Taylor, J. L. S., Rabe, T., McGaw, L. J., Jager, A. K., & Van-Staden, J. (2001). Towards the scientific validation of traditional medicinal plants. *Plant Growth Regulation, 34*, 23–37.

Teng, R. W., Ni, W., Hau, Y., & Chen, C. X. (2003). Two new tirucallane-type triterpenoid saponins from *Sapindus mukorossi*. *Acta Botanica Sinica, 45*, 369–372.

Tepe, B., Daferera, D., Sokmen, A., Sokmen, M., & Polissiou, M. (2005). Antimicrobial and antioxidant activities of the essential oil and various extracts of *Salvia tomentosa* Miller (Lamiaceae). *Food Chemistry, 90*, 333–340.

Terahara, N., Callebaut, A., Ohba, R., Nagata, T., Ohnishi-Kameyama, M., & Suzuki, M. (2001). Acylated anthocyanidin 3-sophoroside-5-glucosides from *Ajuga reptans* flowers and the corresponding cell cultures. *Phytochemistry, 58*, 493–500.

Thalib, A., Widiawati, Y., Hamid, H., Suherman, D., & Sabrani, M. (1996). The effects of saponin from *Sapindus rarak* fruit on rumen microbes and performance of sheep. *Jurnal Ilmu Ternak dan Veteriner (Indonesia), 2*, 17–20.

The Wealth of India. (1972). *A dictionary of Indian raw materials and industrial products. Raw materials, Vol. IX* (pp. 227–229). New Delhi., Rh-So: National Institute of Science Communication, CSIR.

Thomas, J. (1997). *Medicinal and aromatic plants research in India*. In UNDP.

Thorpe, T. A. (1988). In vitro somatic embryogenesis. ISI atlas of sciences. *Animal and Plant Sciences, 1*, 81–88.

Thorpe, T. A., Harry, I. S., & Kumar, P. P. (1991). Application of micropropagation to forestry. In P. C. Debergh & R. H. Zimmerman (Eds.), *Micropropagation technology and application* (pp. 311–336). Dordrecht: Kluwer Academic Publishers.

Tingey, S. V., & Del-Tufo, J. P. (1993). Genetic analysis with random amplified polymorphic DNA markers. *Plant Physiology, 101*, 349–352.

Tripathi, M., & Kumari, N. (2010). Micropropagation of a tropical fruit tree, Spondias mangifera Willd., through direct organogenesis. *Acta Physiologiae Plantarum, 32*, 1011–1015.

Tiwari, P., Singh, D., & Singh, M. M. (2008). Anti-*Trichomonas* activity of *Sapindus* saponins, a candidate for development in microbicidal contraceptive. *The Journal of Antimicrobial Chemotherapy, 62*, 526–534.

Trajtemberg, S. P., Apostolo, N. M., & Fernadez, G. (2006). Calluses of *Cynara cardunculus* var. cardunculus cardoon (Asteraceae): Determination of cynarine and chlorogenic acid by automated high-performance capillary electrophoresis. *In Vitro Cellular & Developmental Biology: Plant, 42*, 534–537.

Tripathi, M., Kumari, N., Rai, N. P., Rai, G. K., & Singh, M. (2012). Monitoring the genetic fidelity of micropropagated plantlets of Spondias mangifera Willd. Using RAPD marker assays. *Journal of Horticulture Science & Biotechnology, 87*, 451–454.

Trojanowska, M. R., Osbourn, A. E., Daniels, M. J., & Threlfall, D. R. (2000). Biosynthesis of avenacins and phytosterols in roots of Avena sativa cv. Image. *Phytochemistry, 54*, 153–164.

Troup, R. S. (1921). *The silviculture of Indian forest trees* (Vol. 1, p. 232). Oxford: Claredon Press.

Tsuzuki, J. K., Svidzinski, T. I. E., Shinobu, C. S., Silva, L. F. A., Rodrigues-filho, E., Diógenes, A. G., Cortez, D. A. G., & Ferreira, I. C. P. (2007). Antifungal activity of the extracts and saponins from *Sapindus saponaria* L. *Annals of the Brazilian Academy of Sciences, 79*, 577–583.

Upadhyay, A., & Singh, D. K. (2011). Molluscicidal activity of *Sapindus mukorossi* and *Terminalia chebula* against the freshwater snail *Lymnaea acuminate*. *Chemosphere, 83*, 468–474.

Upadhyay, A., & Singh, D. K. (2012). Pharmacological effects of *Sapindus mukorossi*. *Revista do Instituto de Medicina Tropical de São Paulo, 54*, 273–280.

Upadhyay, R., Chaurasia, J. K., Tiwari, K. N., & Singh, K. (2013). Comparative antioxidant study of stem and stem induced callus of *Phyllanthus fraternus* webster – An important antiviral and hepatoprotective plant. *Applied Biochemistry and Biotechnology, 171*, 2153–2164.

Upadhyay, R., Chaurasia, J. K., Tiwari, K. N., & Singh, K. (2014). Antioxidant property of aerial parts and root of *Phyllanthus fraternus* webster, an important medicinal plant. *The Scientific World Journal, 2014*, 692392., 7 pp. https://doi.org/10.1155/2014/692392.

USDA. (2010). Sapindus saponaria. Germplasm Resources Information Network (GRIN). Agricultural Research Service (ARS), United States Department of Agriculture (USDA). Retrieved November 2, 2010.

USDA. (2015). Sapindus saponaria. Natural Resources Conservation Service PLANTS Database. United States Department of Agriculture. Retrieved November 5, 2015.

Vanisree, M., & Tsay, H. S. (2004). Plant cell cultures — An alternative and efficient source for the production of biologically important secondary metabolites. *International Journal of Applied Science and Engineering, 2*, 29–48.

Varshney, A., Lakshmikumaran, M., Srivastava, P. S., & Dhawan, V. (2001). Establishment of genetic fidelity of in vitroraised Lilium bulblets through RAPD markers. *In Vitro Cellular and Developmental Biology–Plant, 37*, 227–231.

Vengadesan, G., & Pijut, P. M. (2009). Somatic embryogenesis and plant regeneration of northern red oak (Quercus rubra L.). *Plant Cell Tissue and Organ Culture, 97*, 141–149.

Venkatachalam, L., Sreedhar, R. V., & Bhagyalakshmi, N. (2007). Micropropagation in banana using high levels of cytokinins does not involve any genetic changes as revealed by RAPD and ISSR markers. *Plant Growth Regulation, 51*, 193–205.

Verma, A., & Mehata, M. S. (2016). Controllable synthesis of silver nanoparticles using Neem leaves and their antimicrobial activity. *Journal of Radiation Research and Applied Science, 9*, 109–115.

Verma, N., Amresh, G., Sahu, P. K., Mishra, N., Singh, A. P., & Rao, C. V. (2012). Antihyperglycemic activity, antihyperlipedemic activity, haematological effects and histopathological analysis of *Sapindus mukorossi* Gaertn. fruits in streptozotocin induced diabetic rats. *Asian Pacific Journal of Tropical Medicine, 5*(7), 518–522.

Verpoorte, R., Contin, A., & Memelink, J. (2002). Biotechnology for the production of plant secondary metabolites. *Phytochemistry Reviews, 1*, 13–25.

Vincken, J. P., Heng, L., & Gruppen, H. (2007). Saponins, classification and occurrence in the plant kingdom. *Phytochemistry, 68*, 275–297.

Vines, G. (2004). *Herbal harvests with a future: Towards a sustainable source for medicinal plants*. Plant Life International. www.plantlife.org.uk.

Virdi, G. S. (1982). The piscicidal properties of *Acorus calamus, Sapindus mukorossi* and *Xeromphis spinosa* on 7 species of fishes of North India. *Indian Journal of Natural Sciences, 2*, 28–35.

Vogelsang, K., Schneider, B., & Petersen, M. (2006). Production of rosmarinic acid and a new rosmarinic acid 3-O-beta-D-glucoside in suspension cultures of the hornwort *Anthoceros agrestis* Paton. *Planta, 223*, 369–373.

Von-Arnold, S., Sabala, I., Bozhkov, P., Dyachok, J., & Filomova, L. (2002). Developmental pathway of somatic embryogenesis. *Plant Cell Tissue and Organ Culture, 69*, 233–249.

Vora, R. S. (1989). Seed germination characteristics of selected native plants of the lower Rio Grande Valley, Texas. *Journal of Range Management, 42*, 36–40.

WHO. (2013). *Asthma fact sheet no. 307* (pp. 11–12). Archived from the original on June 29, 2011. Retrieved 3 March 2016. NHLBI Guideline 2007.

Wijeratne, S. S., & Cuppett, S. L. (2007). Potential of rosemary (*Rosmarinus officinalis* L.) diterpenes in preventing lipid hydroperoxide-mediated oxidative stress in Caco-2 cells. *Journal of Agricultural and Food Chemistry, 55*, 1193–1199.

Wikipedia. (2018). Ferry Slik. "Sapindus rarak". Plants of Southeast Asia. Retrieved on 5 June. https://en.wikipedia.org/wiki/Sapindus_rarak

Williams, E. G., & Maheswaran, G. (1986). Somatic embryogenesis: Factors influencing coordinated behaviour of cells as an embryogenic group. *Annals of Botany, 57*, 443–462.

Williams, C. G., & Savolainen, O. (1996). Inbreeding depression in conifers: Implications for using selfing as a breeding strategy. *Forest Sciences, 42*, 102–117.

Wina, E., Muetzel, S., Hoffman, E., Makkar, H. P. S., & Becker, K. (2003). The effect of secondary compounds in forages on the rumen microorganisms quantified by 16S and 18S rRNA. In

Proceedings of international symposium "Application of gene-based technology for improving animal production and health in developing countries" held in Vienna, Austria, pp. 397–410.

Wina, E., Muetzel, S., Hoffman, E., Makkar, H. P. S., & Becker, K. (2005). Saponins containing methanol extract of *Sapindus rarak* affect microbial fermentation, microbial activity and microbial community structure *in vitro*. *Animal Feed Science and Technology, 121*, 159–174.

Wina, E., Muetzel, S., & Becker, K. (2006). The dynamics of major fibrolytic microbes and enzyme activity in the rumen in response to short and long term feeding of *Sapindus rarak* saponins. *Journal of Applied Microbiology, 100*, 114–122.

Winton, L. L. (1970). Shoot and tree production from aspen culture. *American Journal of Botany, 57*, 904–909.

www.efloras.org, *Sapindus rarak* Candolle, Prodr. 1:608. 1824, Flora of China,12:11, retrieved on 17/05/2018.

www.flowersofindia.net/catalog/slides/South%20India%20Soapnut.html

www.ryan-isra.net/lerak-kelerak-werek-rerek-lamuran-manfaat-kegunaan/

Xia, N., & Paul, A. (2012). Gadek. "Sapindus rarak". Flora of China. Missouri Botanical Garden, St. Louis, MO & Harvard University Herbaria, Cambridge, MA. Retrieved 5 June 2012.

Xu, R., Fazio, G. C., & Matsuda, S. P. T. (2004). On the origins of triterpenoid skeletal diversity. *Phytochemistry, 65*, 261–291.

Yao, H. K., Hui, C. H., Li-Ming, Y. K., Ya-Wen, H., Kuo-Hsiung, L., & Fang-Rong, C. (2005). New dammarane-type saponins from the galls of *Sapindus mukorossi*. *Journal of Agricultural and Food Chemistry, 53*, 4722–4727.

Yadav, K., Aggarwal, A., & Singh, N. (2014). Genetic fidelity assessment of micropropagated Spilanthes acmella (L.) Murr. by RAPD markers assay. *Indian Journal of Biotechnology, 13*, 274–277.

Yen, G. C., & Chen, H. Y. (1995). Antioxidant activity of various tea extracts in relation to their anti-mut agenicity. *Journal of Agricultural and Food Chemistry, 43*, 27–32.

Young, I. S., & Wood, J. V. (2001). Antioxidants in health and disease. *Journal of Clinical Pathology, 54*, 176–186.

Yu, L., Zhao, M., Wang, J., Cui, C., Yang, B., Jiang, Y., & Zhao, Q. (2008). Antioxidant, immunomodulatory and anti-breast cancer activities of phenolic extract from pine (*Pinus massoniana* Lamb.) bark. *Innovative Food Science and Emerging Technologies, 9*, 122–128.

Yudha, S. S., Angasa, E., Mariska, R., & Hendri, J. (2013). Water extracts of Sapindus rarak as medium/reduction system for silver nanoparticles formation. *Asian Journal of Chemistry, 25*, 433–435.

Zhu, H., Wang, Y., Liu, Y., Xia, Y., & Tang, T. (2010). Analysis of flavonoids in *Portulaca oleracea* L. by UV–Vis spectrophotometry with comparative study on different extraction technologies. *Food Analytical Methods, 3*, 90–97.

Zietkiewicz, E., Rafalski, A., & Labuda, D. (1994). Genome fingerprinting by simple sequence repeats (SSR)-anchored polymerase chain reaction amplification. *Genomics, 20*, 176–183.

Zikova, N. I., & Krivenchuk, P. E. (1994). Chemical study of flavonoids from the leaves of *Sapindus mukorossi* Gaertn. *Farmatsevtychnyi Zhurnal Article in Ukranian, 25*, 43–45.

Zimmerman, R. H. (1985). Application of tissue culture propagation to woody plants. In R. R. Henke, K. W. Hughes, M. J. Constantin, & A. Hollander (Eds.), *Tissue culture in forestry and agriculture* (pp. 165–177). New York: Plenum Press.

Zimmerman, J. L. (1993). Somatic embryogenesis: A model for early development in higher plants. *Plant Cell, 5*, 1411–1423.

Ziv, M. (1991). Vitrification: Morphological and physiological disorders of *in vitro* plants. In P. C. Debergh & R. H. Zimmerman (Eds.), *Micropropagation, technology and application* (pp. 45–69). Dordrecht: Kluwer Academic Publishers.

Index

R. Singh, B. Sharma, *Biotechnological Advances, Phytochemical Analysis and Ethnomedical Implications of Sapindus species*,
https://doi.org/10.1007/978-981-32-9189-8

Printed by Printforce, the Netherlands